A Chemist's Role
in the Birth
of Atomic Energy

Interviews with
Charles DuBois Coryell
by Joan Bainbridge Safford

A Chemist's Role
in the Birth
of Atomic Energy

Interviews with
Charles DuBois Coryell
by Joan Bainbridge Safford

Edited and annotated by
Julie E. Coryell
for the
Oral History Research Office
Columbia University

Pᑫ
m

Promethium Press

A Chemist's Role in the Birth of Atomic Energy

The interviews and Index of Names recorded for Columbia University Oral History Research Office by Joan B. Safford as *The Reminiscences of Charles DuBois Coryell*, edited by Julie E. Coryell, are published here by agreement with the Trustees of Columbia University.

Jack Aeby's photograph of the Trinity test, 16 July 1945, is reproduced courtesy of the Los Alamos Historical Museum Photo Archives.

Image of President Truman signing the Atomic Energy Act is courtesy of the Research Office of the Harry S. Truman Library.

Image of Cooling Towers Near River is from EcoPics.

Promethium Press
www.promethiumpress.org
Contact: promethiumpress@gmail.com

ISBN- 978-0-9856711-2-9
First Edition July 2012

Charles DuBois Coryell

1912 - 1971

To those who explain science well and to scientific method:
It offers humankind a lingua franca and
hope for understanding and cooperation.
As Charles often quoted, "The truth shall set you free."

Contents

Lowan, Morrison, A. H. Compton; psychological depression; associations of science workers, Army relations; principles and ethics; Turkevich, move to Oak Ridge with Levy, Cohn, Cole, P. C. Tompkins, E. R. Tompkins; DuPont relations and Quaker democracy; Hanford; construction of the hot laboratory; AUDIO tape begins; safety; Stang; first chemical production of barium, lanthanum; Oppenheimer; CDC's bet; curies trucked to Los Alamos, 18 September 1944, Doan's Ode; RaLa (radiolanthanum); promethium discovery

The Audio File (27.06 minutes) is available for the last
quarter of Session Three, pp. 125 – 141.

Original Preface

This memoir is the result of a series of tape-recorded interviews conducted by Mrs. Joan Bainbridge Safford for the Oral History Research Office with Dr. Charles Coryell in Cambridge [and Lexington], Massachusetts in 1960.

It is one of a series in which the subject was unable to edit his work, and it has therefore been submitted in this form, with only minor corrections and proper name clarifications. The reader is enjoined to bear this in mind, and also to remember that he is reading a transcript of the spoken rather than the written word.

This manuscript is closed until the death of Dr. Coryell. Thereafter it may be read, quoted from and cited by serious research scholars accredited for purposes of research by Columbia University, in such place as the University may provide. No reproduction of this memoir, either in whole or in part, may be made by micro-photo, typewriter, photostat, or any other device, except by Dr. Coryell's heirs, legal representatives or assigns.

Daughter's Preface

In the company of Michael Peter Goldring, in the spring of 2007, I visited the Oral History Collection of Columbia University to undertake the editing task Charles Coryell would have wanted and to make available the one surviving audio reel. My father was Chief of the Section on Fission Products first at Chicago and then at Oak Ridge. He and members of his group designed and used the world's first "hot laboratory." As the audio file records, they exceeded the initial specification for one curie of radioactivity and dramatically surpassed the request from Los Alamos to separate 100 curies of barium-140, by shipping more than two and a half times that amount on Barium Day, September 18, 1944. Later, J. A. Marinsky and L. E. Glendenin discovered element 61 and appealed to my mother, Grace Mary Coryell, to name it. Despite the super secrecy, she grasped the onset of the atomic age, "Name it after Prometheus."

With his interviewer, Joan Bainbridge Safford, herself a child of Los Alamos, Radcliffe College graduate, and intern for the Columbia Oral History Project the summer of 1960, Coryell stated, "I can visualize the history that you want to be composed of three parts. One, a technical history of the science of the Manhattan Project. The second, a personal history of how people were related to one another and how ideas were produced. And the third, how scientists had a political orientation, what factors guided their political orientation and their political initiative."

Because the surviving typescript on onionskin-like paper with variable ink hues proved transmissible only as a PDF, successive assistant curators, Courtney D. Smith, Rosemary E. Newnham, and Corie Trancho-Robie advised me in preparing the corrections by hand. My son, spouse and son-in-law, Carl Coryell-Martin, Seelye Martin and Darin Reid, together with Matt Quarterman taught me computer skills and ergonomics. Paul Wolman helped assemble the work. My half sister, Pat Huber, and stepbrother, Edwin Buchman, generously shared primary sources. Sharon Beck cheered me faithfully as did David and Lois Madsen. Alice Ingraham Davies proofread to Charles' standard. Cary Cartmill prepared the photographs. John Oberteuffer gracefully guided me through publication of the manuscript. My profound thanks to all.

Along the way of editing, Coryell's students and colleagues shared memories, documents, explanations, encouragement. I thank Ralph Cicerone, William and Barbara Walters, Lionel Goldring, Howard Gest, Donald Wiles, David and Linda Freeman, John Winchester, Jack Siegel, William Zoller, Namik Aras, David Samuel, and Israel Dostrovsky. Three: Lionel, Howard, and Jack, were members of the Fission Products Group and participants in Barium Day at Oak Ridge. Especially I am grateful to have rediscovered Joan Bainbridge Safford.

A scholarly work is founded upon dedicated librarians and archivists. I thank Felicity Pors who, at the start of my work, welcomed me into the Niels Bohr Institute Archives at the University of Copenhagen. I thank particularly Andrea Twiss-Brooks, Co-director, Science Libraries Division, Crerar

Library, University of Chicago, Loma Karklins, Archives Assistant, California Institute of Technology, Amy McDonald, Archives Assistant, Duke University, Myles Crowley and Frank Conahan, Institute Archives and Museum, and Elizabeth McGrath, Communications and Development Director, Department of Chemistry, Massachusetts Institute of Technology. At the University of Washington, Physics-Astronomy Reading Room director Pamela Yorks and especially Chemistry librarian Susanne Redalje extended themselves to my benefit.

Blessed with an amazing memory, storied creativity, generosity, Coryell defined himself as emotional and prone to boom-bust phases. In these reminiscences he relates "alarms and rebellions," war-defined urgencies, risks he willingly "bet on a nickel." In college, he renounced anti-Semitism, became bilingual in Germany in 1933, and from A. A. Noyes, Linus C. Pauling, James Franck and others learned how to conduct research. He forged a high personal integrity tempered by pressures to win World War II and prosper peace. I offer these corrections and few notes wherever possible with humility and love for his scientific and familial legatees. He used to say, "Doubt yourself," "Cudgel your memory," and "Nobody likes negative results, but they are results." Not meant as discouragements, but as incentives to do your best, and reach beyond yourself. Now the fission products, daughters, speak not only of creativity and chemistry, but also of connection and universality.

Julie Esther Coryell

Session One — 14 July 1960

Coryell: To start with, my name is Charles D. [DuBois] Coryell. I was born [February 21] in 1912 in Los Angeles, California. I had education in the public schools, principally Garvey Grammar School, Alhambra High School, and I decided to become a scientist my junior year of high school and found I hadn't taken the right program to get to Caltech [California Institute of Technology, Pasadena], where I was pressed to go by my high school chemistry teacher, Mr. Bragg. So I took a year off to make a little money, working in a bank at sixty dollars a month and going to night school downtown, Manual Arts High School. I entered Caltech in September 1929, with a scholarship. The biggest thrill of my life was to be in college, and I am very pleased that I got into Caltech. It was a very small school of very high caliber, and by the end of the first year I knew something about most of the members of the Chemistry Department. Classes were a hundred and sixty in freshman, eight sections of twenty each. I was from the second quarter on in the honors section, and I had a most stimulating undergraduate program.

Caltech, like Stanford and Chicago, sprung full-armed from the brow of Zeus. It was founded [1891 as Throop College, renamed in 1921] about 1920 by A[rthur] A[mos] Noyes

(1866-1936), with whom I later took my PhD, who was someone I never knew [before], and Robert Andrews Millikan (1868-1953), who's of course, very well known. They geared for high standards, and I realize now, sitting at MIT [Massachusetts Institute of Technology], that Caltech designed itself much in the pattern of MIT. This is rather widely recognized now with the great friendship existing between the two institutions. [Noyes, professor of chemistry at Caltech 1919-36, served as acting president of MIT, 1907-9. From the mid-1960s Charles named his laboratory at MIT for A. A. Noyes.]

In the student body, there was a strong feeling of rivalry between the scientists and the engineers, and I was in the science group. I planned to be a chemist without knowing much what chemistry involved for a living. In fact, I specified I was going to be a chemical engineer, but emotionally from the start I was with the science group. The chemical engineering was a small part of Caltech in those days. Only a Master's degree was offered. In my freshman year, Noyes sent a message to me and the other students registered as chemical engineers and asked us to consider taking training in chemistry and doing the engineering later on. I acceded to this. I also won a summer scholarship to work at Caltech's [Kerckhoff] Marine Laboratory [Corona del Mar], at the end of my freshman year. The five best freshmen according to a test examination in chemistry were selected to have a summer scholarship, and we did our whole sophomore chemistry there.

It was a trick to get the bright boys into chemistry from physics. Of the five going down, I was the only chemist, and of the five coming back, I was the only chemist. Caltech also had an undergraduate travel scholarship program for juniors. The eight best juniors were given special training in languages, as well as the art and architecture of Europe, and a prize. Two nine hundred-dollar prizes were offered at the end of the second quarter of junior year. The young people took the third quarter and had a trip to Europe. I was the winner in what must have been January of 1932, but I was married to a childhood friend and high school companion [Meta Seward (1912-1989)] secretly in the fall of '3[0] [December 6], and I didn't want to take this European trip and lengthen my bachelor's program, because, as a result of having been to the Marine Laboratory at Corona del Mar, I'd gotten a year ahead in chemistry and was prepared to graduate in three years from Caltech. So I stayed and did that, took my travel prize in the following year. To have taken the prize would have cut me out of a three-year graduation, and times were hard, and I wanted to get ahead. I had a wife and a child [Patricia Louise Coryell Bowler Huber (1931-)].

A. A. Noyes, it later developed, gave this prize money out of his own pocket. It was a secret award—a secret source of funds [patent income, Hager 1995]—[and he] also loaned me enough money so that I took my wife to Europe. He also felt that young men should have, as he had had in the late 1880s, a year in Germany if they were very promising. In the fall of 1932, he supported me for a German-American exchange fellowship. I received this, but these were normally not given to married people, and I received it under the specification that my wife not be with me during the tenure of the fellowship. So I combined the travel prize with the one from Caltech to take my wife to Europe in June of '33, and then she came back home the first of November with our joint passport, and mailed it back to me, and then I stayed on until the following July.

It was the *Technische Hochschule* in Munich. By having delayed this trip, I accomplished a first graduate year before I left. And this, therefore, did not impede my getting my PhD, because having good graduate status at Caltech before I left, I could use my German material as part of my thesis. The middle portion of my thesis is in German. And then I had one more year back at Caltech, and I got my PhD in May 1935.

I also, on returning from Germany, found that Noyes wanted me to do my graduate work with him. I had been on a problem I didn't do very well on and wasn't very happy about, for technical reasons and probably emotional shortcomings of my own. The heart of the problem related to photochemistry, and when I returned, Noyes grabbed me for a job on silver, and I had the great honor of being his only doctoral student the last year of his life, and I got to know him personally very, very well.

As far as I can tell, my scientific position has been patterned after the model offered me by A. A. Noyes, Linus [Carl] Pauling (1901-1994), also at Caltech, and James Franck (1882-1964), who was for a very short period of time my direct boss in the Manhattan Project during the war. And these three men are substantially different, one from the other, and I sort of feel I would like to combine important parts of the temperament and direction of thinking of these three people. So a few more words about Noyes is probably in order.

Noyes was the [professor] I got to know rather well during that first summer when I was a freshman. Noyes was very anxious to have freshmen begin research, and all the honors students, forty out of the hundred and sixty of us at Caltech, no, twenty of the hundred and sixty of us did research problems in their first year. In this capacity I worked with [Arnold Orville] Beckman (1900-2004), who is now a famous producer of scientific instruments in California. He was a very brilliant young man, also a protégé of Noyes. And Linus Pauling, himself, was a discovery of Noyes. Noyes selected roughly one man out of each class to give a fair amount of personal attention to, and otherwise was not known to the students. He never gave a public lecture or a scientific lecture that I heard. He was ill part of the time and was a lonely, very modest quiet person. But he formed very close friendships with this group of young men, and I was in the last two or three of these, because he died in June of 1936.

The two others that I was close to—knew well—were Donald Peter Bolt, who went to Berkeley for his PhD and got into nuclear science with [Willard F.] Libby (1908-1980), was a conscientious objector on religious grounds during the war and spent some time in prison. And in the late '40's [he] left science as a research field, and was teaching at the College of the Pacific [University of the Pacific]. And the other one, who dates slightly after me is Richard W. Dodson (1915-2002) [BS 1936], who went into nuclear chemistry for his PhD at Johns Hopkins [1939], and was prominent at Los Alamos. He's now chairman of chemistry at Brookhaven [Brookhaven National Laboratory, Upton, New York, 1947-]. Like me, both of these men were very much influenced by Noyes.

Dodson, with respect to politics, was totally apolitical except when the [J. Robert] Oppenheimer (1904-1967) case came up [April 12, 1954]. And his feelings were greatly involved, and he was strongly and courageously pro-Oppenheimer at a time when it was pretty risky. He'd been secretary of the General Advisory Committee to the Atomic Energy Commission [Atomic Energy Act, 1946 - Energy Reorganization Act of 1974], though I think he didn't testify. But it is the only point where he and I have a common front on science and politics, or scientists and politics.

Well, as a result of the separation in Germany with my first wife and family pressures of various sorts, she sued for divorce on non-support, and I got the subpoena for the uncontested divorce the day I took my doctor's exam at Caltech.

Also the spring of '35, the divorce was coming up, Pauling asked if I would join his group, be the first person to work on the money he had from the Rockefeller Foundation for the application of structural chemistry to [biology] and before I got my PhD, as soon as I'd gotten my thesis accredited, I started work with Pauling on magnetic problems in the structure of hemoglobin compound, and I worked three years post-doctoral with Pauling in this field. I worked in physical biochemistry, principally hemoglobin family compounds, at UCLA [University of California, Los Angeles], until I was opted into the Manhattan Project and dropped the biological nucleus for the atomic nucleus.

I worked very closely with Pauling these three years; I'd known him before. In fact I think I'll cut back and tell a little story in this connection which will cast a light on the whole Caltech relation.

Noyes was strong for research for all undergraduates, as far back as the freshman year for bright undergraduates, and this affected my life a great deal and I feel strongly the same way. The problem that I did with Beckman—Beckman offered a set of problems and we paired off, I went off with a young man named Moses Verder, whom I want you to know I haven't seen since I left Caltech, and we worked on a small problem involving hydration of cobalt chloride into solid, cobalt chloride hydrate. But I learned of some of the attitudes of research, and some of the frustrations of research, and some of the pleasures.

The work at the Marine Laboratory at Caltech was under E[rnest] H[aywood] Swift (1913-2002), who is now Chairman of the Chemistry Department at Caltech, and we did a little bit of routine quantitative analysis, and then did research problems. The various things we worked on and other students came out in the major book on Swift's life on the system of qualitative analysis. [*The Use of Quantitative and Qualitative Analysis in the General Chemistry Laboratory*, 1959.]

Noyes was also in the lab to some extent and had a couple of graduate students working with him, that summer of my freshman year, 1930. I also should say that Noyes was revising the major book that he had in physical chemistry, *Chemical Principles*, by Noyes and Sherrill. [*A Course of Study in Chemical Principles*, June 1939]. [Miles Standish] Sherrill (1888-1965), was a professor of chemistry at MIT, in residence at Corona del Mar, near Balboa, California, and came over every Friday to a seminar with the five of us

7

students, A. A. Noyes, Miles S. Sherrill, and E. H. Swift. And so I got another lead on thought development in chemistry.

Well, when I came back to the start of what was my second year at Caltech, Noyes thought I should start a real research problem and he wanted me to talk with Pauling. Well, Pauling was friendly, but asked how much glass blowing I knew and could I work in the machine shop, and I could do no glass blowing to speak of and didn't know one end of a lathe from another. So Pauling politely passed. Pauling said that every young man should do an electrode potential and he sent me to work with, to see Don M. Yost (1893-1977). And Yost accepted me, and so in my sophomore year I started a research problem, which wasn't very spectacular either. It was the possibility of making the compound iodine monofluoride. It's not yet been made and probably doesn't have any existence. But I went further in my stumbling for a research point of view. And I did another job the following year with Don M. Yost, which was an electrode potential, and this led to my first publication. [Charles adopted Yost's warm western informality, "Call me Don."]

The publication, the work was finished the summer following my third year; June of my third year was the year that I got my bachelor degree with honor. I guess I was the second man in the history of Caltech to go through in three years, but the history of Caltech isn't a very long one, and it's been done a lot, quite a bit since, probably. Well then, in the fall [and] in the summer following my graduation [B. S. 1932], I worked on the problem with Yost. And in the fall I started out with [Roscoe Gilkey] Dickinson (1894-1945). As far as I was concerned, the problem was abortive. Then

8

Noyes used me some for emergency work in the spring of '33 and then in the summer of '33 I went to Europe.

And in Europe I worked on a photochemical problem with negative results, fluorescence, the search for discreet fluorescence of acetone. But the report I left in the laboratory in Germany, I wrote my own bad German, [which] I doctored up to be a little more formal German, changing present to past, and checked with a German-born friend. It is the middle part of my thesis in German at Caltech.

Then in '35, starting in May, I was working for the princely sum of $1500 a year for Pauling and I've never had a more happy scientific life than in the three years beginning then. I think not that my life with Noyes was less satisfying, but I was full-time research with a much more frequent personal participation of my research supervisor. Noyes was reserved, with a deep warmth, but Pauling was warm and effervescent, with a good sense of humor and treated me as a younger brother. I also got to know the Pauling family. I babysat for them, and did all the things that Caltech post-doctorals did.

And there was also a very stimulating overall scientific life, because by then I knew all the staff. We were friendly with the graduate students, but we post-doctorals had more nerve to ask questions in seminars and we got really the best of intellectual life at Caltech.

I better stop, though, and make some broader comments. I may say in the way of passing that my students here at MIT gave me a Doctor of Digression as this tape will show before we're through.

I can visualize the history that you want to be composed of three parts. One would be a technical history of the science of the Manhattan Project. The second would be sort of a personal history of how people were related to one another and how ideas [were] produced, but with not great technical detail, related to the Manhattan Project and the things that have come out of it. And the third would be how the scientists had a political orientation, what factors guided their political orientation and their political initiative. What I talk about is not very chemical. It's sort of pre-Manhattan Project, presumably background, of what I'm going on into.

Well, in the framework of part three, the political orientation, I come from a family which was Democrat since the day of Bull Moose, and lower middle-class, and rather hard up during the depression of '23 and again during the depression from '32. The depression didn't hit California very badly until '31, but my family was never very well off when there wasn't a depression. And going to Caltech was some sacrifice, though I paid my own way, tuition-wise. I think I may have had half-tuition the first year. The little savings I had [I] made in this stinky job in the bank. I went back to Alhambra High School, also the second semester between high school and college, and took physics, trigonometry, and solid geometry, and worked in a grocery store and vegetable stand. So there was a little bit of money in the family to take care of that. But my family without any, sacrificed more for me than I realized until later on, because they believed firmly in education. My mother [Florence Elizabeth Cook Coryell (1881-1947)] had been a schoolteacher. I think her influence was predominant. There was never any objection to this shortage of money, matter of fact, my family was much poorer than I realized, we never thought about money. The money always showed up at tuition time, and I did my share to keep grades up and earning scholarships.

The family wasn't very sophisticated about politics. We read the *Los Angeles Times,* which wasn't as bad then as now. My father [William Harlan Coryell (1884-1953)] was very anti-Semitic; my step-grandfather [m. Lucia Lou Coryell], far more so. My father and mother were both from Omaha, Nebraska. I didn't know any Jews. The first Jew I got to know well when I was about thirteen or fourteen was an awfully warm-hearted little guy with a father and mother with thick accent who lived back of a small store in our neighborhood. This young man was so much different from the picture I'd heard of the dirty Jew, that I immediately decided that I couldn't follow my father's point of view and became more interested in finding out more about Jews. The first time I met many was in my freshman class at Caltech. I would guess now that the number might be ten percent. And I've been pro-Semitic ever since. Many years later, James Franck, no, Nathan Sugarman (1917-1990) and Anthony Turkevich (1916-2002), Turkevich being a son of an Archbishop Metropolitan of the Russian Orthodox Church, made me an honorary Jew in joking response to my defending Jews because I thought the Army was discriminating against them.

The first use of the "gray list," charging guilt by association, was in 1944; when the Army was cutting down the Project, it was indicating what names to remove first. The story went around that these were a "gray list." These were people who were personally reliable, but whose associates were not. And the story that came to me from Sugarman, Turkevich, and some of the people involved, Turkevich especially, was that Farrington Daniels (1889-1972) was told that, if he dropped these people gently, but helped take care of jobs, that the Army wouldn't use this trick anymore. As long as they were going to cut down personnel, they might as well cut down that personnel which was most worrisome to them.

But it happened that except for two Negroes everybody on that list that I know was Jewish. It seemed to me that it was marked discrimination involved, and not being Jewish I could say what I thought about the Jewish matter, and I could speak more frankly being a prominent person at Chicago who was based in Oak Ridge [Tennessee], than the people who had to live with the problem all the time in Chicago. And on that basis I was made an honorary Jew.

Since then I have had a fellowship in Israel [Louis Lipsky Fellowship at the Weizmann Institute of Science, Rehovot, 1954] and have been very much interested in the relation of the Jewish intellectual community in the world to the world and to Israel and the situation of the American-born children of Jewish immigrants, the family conflicts, [the] part of straddling of cultures.

The family wasn't very politically oriented outside of being die-hard Democrats. I remember, it must have been the election of 1920, that I was the only kid in the neighborhood whose father voted for Davis. I remember being razzed by the kids whose fathers and mothers voted for Harding.

When I came to Caltech, there was a systematic effort on the part of the faculty to point up for us young people all of the world problems. The idea was to make the scientist and the engineer take the same position in decision-making of the world that the lawyers traditionally have. We had eminent writers and philosophers speaking at the Monday or Friday assemblies, I've forgotten which day it was that Caltech had. And we were sort of given the impression that it's our responsibility as among the intellectual élite to be responsible in all areas of human endeavor.

Later on, as the Spanish war began to loom, well, matter of fact before this, I realized when I got to Germany as an exchange student in the summer of 1933, I had no detailed knowledge of what the Nazi movement involved. I had been reading *Time* ever since current affairs days at Caltech, my third year there, and I had a pretty good knowledge of general European history, and I had been quite a student of World War I, and I wrote a term paper on the Reformation and things like this. I was fairly religious Episcopalian in those days. I was a Sunday school superintendent until I graduated from high school. But I was sort of a babe-in-the-woods in the German atmosphere. But it didn't take me very long to find out one thing for sure, Germany was girding for war, psychologically as well as physically. And there'd be an awful war in Europe. I couldn't tell whether it would be two years or ten years away. But Germany was going to play a prominent part.

I was not at all, I mean I was opposed to the Nazi position against Jews. I knew a few Jews in Germany. But I didn't realize the magnitude of the cruelties. The cruelties I knew about were economic discrimination. Hitler in *Mein Kampf* in 1923-24, outlined his position about the Jews, but the Hitler government came to power in January, 1933. And my

13

fellowship for Germany was arranged or applied for before this. And I came into Germany in August of '33 with my wife from Louvain [Belgium], where we had seen the library to repair the one burned by the German hordes. There is some expression in Latin that has to do with the German hordes. [Charles studied Latin in high school. His mother studied Greek and Latin at college.]

It happened to be the day that Germany voted to withdraw from the League of Nations, and the station at Köln was loaded with young kids in brown clothing singing songs of German nationalism, and on the street the expression was, [blank], "Day of the Geneva No." And Germany voted about ninety-nine percent to withdraw from the League of Nations. [November 12, 1933, 93% voted to withdraw, and 92%, in favor of National Socialist candidates.]

I was in Munich during the blood purge. I was naïve enough not to have much prejudice except to believe that it was dreadful to be anti-Semitic and to be able to talk with Germans I met freely, but not to carry on a crusade and not to realize how dreadful the situation was going to be. In this particular case, the date of the [Ernst] Röhm arrest [June 30, 1934], there was marching in the streets about two o'clock and singing and the next morning there were no newspapers. The Germans were all standing about in knots talking with one another and didn't particularly want me to participate. I was in the group from the *Technische Hochschule* Institute of about twenty Germans where there was only one other foreigner besides myself, an Italian.

But then they found out that I could get foreign newspapers at the railroad station and they couldn't get them. They didn't dare get them; they were afraid of getting into trouble. So I picked up the *New York Times* from Paris, no, the *New York Herald-Tribune* from Paris and the *London Times*, to get a little news about the uprising. This was probably a Friday morning. [The purge ended early Monday morning, July 2, 1934.] It wasn't until Sunday that there was an official German statement. Röhm was taken prisoner and then later on shot.

I heard from the Germans that there were soldiers with machine guns down at the Brown House, the headquarters of the Nazi Party in the [Briennerstrasse 45]. So I immediately go down to see. The Treaty of Versailles forbad Germany to have machine guns. So I stepped up to ask a soldier, "What are the machine guns for?" He said there was an important meeting in the Brown House. He might have knocked me down. But I guess he felt that with my accent and clothing that I was an American.

And then, I followed the newspapers. When I bought a newspaper at the station, some old German came to speak to me and told me how terrible things were. Apparently he believed the Nazi propaganda about things in general. He had been reading a lot of stuff Röhm had said. And then he was so horrified that a man who had written this wonderful book that he had just gotten out of the library, should have been put to death so suddenly. I mean he realized that the arbitrary nature of the power shifts and presumably the arbitrary basis on which they stand. The man who believed in fundamental principles, not the man who believed in the opportunism ...

But I observed lots of things in Germany, which I later could interpret much better, because after I got home I read much more widely on the course of things in Europe.

But I think one observation that's probably worth making now related to the problem of Jewishness. As nearly as I can tell a large fraction of the people accused of being spies, accused of being disloyal, and the people who have been established to have been subversive in their activities are Jewish. And I know this worries people in Israel and worries conscientious Jewish people a great deal. It is more than the anti-Semitism I spoke about in the Army action. But I think that the story is probably roughly this, that in the mid-thirties, Jewish people knew from personal correspondence and contacts what was going on in Germany, and most non-Jewish people had no realization of how cruel things were or didn't move very much or just didn't accept it. So the Jewish people who had friends in central Europe were more articulate, joined organizations and often these were communist-infiltrated. After all, the Soviet government had the same nominal position, idealism about race and all, that we in America had. Then later on some of these people were trapped in some of these positions or at least blackened by these associations which they had made in a perfectly natural fashion. This seems to have been the case with Oppenheimer.

He realized during the Spanish War, and it's also true that Jewishness was involved with liberalism, and he went over-liberal, and then withdrew a bit. And some of these cases like the [Wendell H.] Furry (1907-1984) case [at Harvard] and the [Isadore] Amdur (1910-1970) case at MIT are probably this way.

16

Well, I had emotional warning from my German stay, and on coming home I became more and more interested in world affairs. I guess I became socialistically minded ever since my sophomore days at Caltech. William B[ennett] Munro (1875-1957), who founded the Harvard School of Business Administration [CDC error: Munro taught government and history courses at Harvard, 1904-1929], spent the last and rich portion of his life at Caltech [Professor Emeritus of History and Government, 1929-1945, served on Executive Council and Board of Trustees]. He gave the history lectures. He was a backbone of Republicanism but, when the lectures came to discussing anarchism, socialism, and communism, he gave a very good presentation of the case and convinced me of the merits of socialism in a small course in sophomore year.

In 1932, I wasn't quite old enough to vote, but I campaigned for Norman Thomas and, remember, the Depression had hit California by then. I remember attending Pasadena Junior College, a Thomas talk, and the college auditorium was full, and about half the Caltech staff was there. They had to put chairs on the stage. It was just loaded with Caltech staff and I was very proud of them.

Well, as the Spanish Civil War broke on the scene, my sympathies were totally with the Loyalist government. I felt that this was the first stage of a major war in Europe. And I had very deep emotions about this. I was, however, involved in the divorce situation and the like, and very short of money. So I had strong sympathies for the Spanish government. I gave fifty cents now and again to supports of various sorts, but I wasn't emotionally constituted to go off and fight in a Loyalist brigade, but I wasn't emotionally opposed to it.

Some of the meetings that went on at Caltech at this period developed from strong sympathies for what I called the "right side" to groups that were communist-captured. I joined in 1937, probably, a social study group, and we went through Strachey's struggle for power. And I found this quite stimulating to discuss the theoretical aspects of essentially pretty socialist, Fabian socialism. Apparently, some of the people in that group later on went a little more directly into communist activities. One of the members of this group was [Joseph W.] Weinberg [*sic*] [probably, Sidney Weinbaum, Pauling research associate, Hager 1995], who was a Russian-born man around Caltech. He was very sympathetic with communist causes, though I think he was so warm-hearted that he may never exactly have held a Communist card. He raised funds for the Spanish War, and he came to collect money for the Communist Party. He never tried to get me to come any closer and I knew him quite well as one of the old stalwarts of the laboratory. He was later indicted for falsely reporting that he was not a Communist on government papers during the war, and was tried for perjury, and got a jail sentence in the late '40s. Pauling got his fingers very badly burned in this case.

The reason I'm giving you a little bit of the background is that in this period, I was occupied internally by [my] family situation. Also after my divorce became final in 1936, I met about six weeks later Grace Mary Seeley (1914-1965) from a very conservative family. She disapproved highly of my spending time and energy on this kind of stuff when there was either science to do or taking care of the family things.

But there was a lot of feeling among many of us including me, about Pauling, that there was something wrong with him that he showed no interest in this sort of thing. Like [Enrico] Fermi (1901-1954) was most of his life, he was completely apolitical. He was just too busy in science, too busy with people to be concerned about the general theoretical problems of the world.

The first time I've known him to take any public position in science was in 1940 at one of the Caltech student assemblies. He spoke on behalf of [Franklin Delano] Roosevelt (1882-1945), and one of the members of the economics department, Horace N. Gilbert, spoke on behalf of [Wendell L.] Wilkie. These were just presentations to the student body of positions. The student body was polled just about that time and disgusted lots of us by being about sixty percent Republican, or seventy percent. I liked the man personally, I almost voted for him. But instead I voted for Norman Thomas. The move for Wilkie was a move against the formal politicians for the sentiment of the people. And I think, as far as I know, Wilkie honestly reflected a warmth and a popular interest that wasn't characteristic of his banking background. He might have made a very good president. At any rate, in 1932 I couldn't vote, but I was strong for Thomas. In 1936, my wife's uncle was so sure that [Alfred M.] Landon was going to win, that I abandoned my principles and voted for Roosevelt, and then was disgusted ever since that I had wasted a good socialist vote, because I always like to vote with the minority to the point where the majority ought to look. In 1940, I felt sure Roosevelt would win, I preferred, I didn't know whether a third term wasn't a little too much, but I also didn't trust Wilkie's background. But I was also very strongly against the war, so I voted for Norman

Thomas again. In 1944, I couldn't stand [Thomas E.] Dewey, and we were in the middle of a war, and Roosevelt represented a man who knew something about atomic energy, and I would do everything possible to see him and his principles continued in office.

But he didn't pass his feeling for atomic energy on to anybody. He was scheduled to visit Oak Ridge the week that he died [April 12, 1945]. And we were looking forward to his visit though I suppose we'd have seen him from a distance. In his place, [Secretary of War, Henry Lewis] Stimson (1867-1950) came with the second general in the atomic energy project after General [Leslie R.] Groves (1896-1970). [Cf. Nichols 1987, 164-7.]

I have described in fair detail some of the forces in my undergraduate life, and my three year's post-doctoral work with Pauling, and the principal aspects of my German experience, which were mainly to make me strongly in favor of exchange-student arrangements, for students who know enough to look at things with their eyes open. I got a fair amount of science done in Germany, but I got a large lovely cultural improvement. Because I came from an uncultured, well, not an uncultured, not a very sophisticated cultural background.

JBS: You said that when you were in Germany you have been aware or became aware of the Nazi threat.

Coryell: Well, I recognized the Nazi threat, principally in Germany. It was obvious that it was a march toward war, which Germany would be able to withstand the pressures of very, very effectively. I had little personal evidence about

Nazi cruelty. And I had what I could read in *Time* magazine. *Time* magazine finally got censured in Germany and they wrote saying that they couldn't guarantee my receiving my subscription. So I was fool enough to say, "Okay, don't send it." It turned out that it did come to Germany most of the time. But when I got home I read all of the copies of *Time* back up. Then I read about Europe with open eyes.

I also was extremely interested in the process of becoming bilingual. This had a big effect on me and was something which was a great delight. And it has been a great advantage to me to have a foreign education added to my American education, from the personal point of view and the cultural point of view, and I think it's a wonderful thing to be multilingual. And I think that it's a wonderful thing to have international connections, international experience. I have done everything possible to understand foreign students, to generalize the experience I got, to help other people do it, get fellowships, and help handle other foreign students that come in here.

JBS: You said you voted for Thomas rather than Roosevelt because of the threat of war. How did you reconcile your feelings about Germany with this pacifism?

CDC: I was very depressed. I was depressed already in 1932 when the United States couldn't get any of its friends to do anything about Japan and China. I was very depressed when England step by step—when Europe—didn't take any strong action about the Saar, when refugees from Austria, when I began to meet refugee scientists coming to California. Men like Otto Redlich (1896-1978), who was a year or two finding a decent job—the best physical chemist of central Europe.

Then when Germany sold, took part of Czech Sudetenland, this was very disgusting, and then the [Neville] Chamberlain compromise [Munich Pact, September 1938] and the final loss of Czechoslovakia. A friend of mine whom I think is also quite liberal, Jim Gilluly (1896-1980), made the clever statement when war broke out in 1939, "I have just as much sympathy for England as England had for Czechoslovakia."

I felt to a fair degree the socialist argument, that this was just two imperialist powers neutralizing each other, rather than any fundamental principles which was largely right. And I didn't want to see the United States sucked into this sort of war in Europe. But then as the war developed from [blank], and also the true stories of the atrocities came home to me, then I became very much concerned and I felt also we were most likely to get into the war by an attack in the Far East. Already in the 1930s, I wouldn't let my wife buy Japanese silk, and I objected to selling scrap iron to Japan. I was worried about Fascism on a general scale then. But I didn't see the start of the war in Europe, that it was going to solve the problems very directly, and I didn't want to see the United States go into this war unprepared. I can't analyze fully now my whole feelings about this.

JBS: To what extent was the Caltech faculty involved in the Spanish cause?

Coryell: Well, a number of younger people, some of whom since have been identified as Communist allies, were very active in raising funds. I don't know personally any man who went to bear guns. And some of the staff, Henry Borsook [Professor] in Biology [Division, 1929-1968], for instance, was very active in teas, art functions, and the like, to raise funds for the Spanish War. I think I attended one of these, and I often gave fifty cents or a dollar when they were collecting funds. Every time there was a full moon I would think, "Well, this will be a hell of a night for Madrid." My wife used to be horrified by the way I'd keep bringing up the dark thoughts of war when everything was very pleasant in California. [These included] the bombing of Madrid, the whole destruction of the Loyalist position and what I gather the cruelty of the Franco position, and the strong participation of Germany and Italy on the Spanish side with nothing but the Russians on the Loyalist side. I also had some suspicions about the honesty of the Loyalist position.

Now when the first Finnish war came along, I had tremendous sympathies for Finland, but I had—I think I said to people—I want to see what happens. The situation in Europe is so full of war threat, that I can understand a reasonable basis in self defense of Russia taking the Hanko Peninsula and taking the [blank] isthmus, if they go no further than this. Though I was delighted when the Finns seemed to lick the Russians for some time. I must say in this regard, that it wasn't too much later that German forces came through Finland to attack Russia. And Leningrad was surrounded for a year and a half. And at the end of World War II, Great Britain declared war on Finland and the United States did not, and that at the end of World War II, Russia withdrew from most of Finland and a few years ago they withdrew everything from Finland except the [Hanko] Peninsula through which they came.

Now this dividing of the [?] Finnish war is one of the principles one uses to decide how pro-Communist people were. I wasn't a part, I had no contact with the Party line. As

I say, nobody ever worked hard on me to make a Communist out of me. My only participation with these people were in open and free discussions in the laboratory, where my position was somewhat ambivalent about lots of these things, both about sympathy with Russia in the big, but worry about the honesty and decency inside Russia. And problems like worrying about the Allies, and feeling that if the United States got involved in a European war, that the honesty of England and France and the United States would be no greater with respect to the 1939 war as it was with respect to the *cordon sanitaire* and the nominal neutrality in the 1937 war. But I don't think that I was strong for the neutrality principle. I didn't oppose conscription, though I wasn't terribly much in favor of it and I didn't like required ROTC [Reserve Officer Training Corps].

I remember going with the wife of one of my friends to an anti-war rally about four months after the war began at UCLA [University of California, Los Angeles], across the street from UCLA. UCLA wouldn't let it appear on the campus. The *Los Angeles Times* took a photograph of the crowd and I was amused to see myself in the *Los Angeles Times* photo section.

Well, this presumably takes us scientifically until I went to UCLA in 1938, and politically, roughly through the end of '39, before the war got mean in Europe.

Session Two — 23 July 1960
(Lexington, Massachusetts)

Coryell: I had one series of contacts involving a small school in California called Deep Springs, a junior college, in fact, that gave me some insight into character formation in students, and in myself probably, and gave me interesting contacts with several people who later had some part in the Manhattan Project.

Deep Springs is a [two-year] college, founded about 1919 [1917] in its present position, by L. L. Nunn (1853-1925), one of the robber barons of the late nineteenth century, who built the first high tension power line in Colorado to bring power to the mines of Telluride (12,000 feet altitude) so they could be operated in winter. He was anti-woman, and believed that the way to develop character was to give responsibility and fight with the boys. He founded the school briefly near Jamestown, Virginia, and it was too close to civilization. He moved it to its present position in Deep Springs Valley [Inyo County], about thirty [forty-five] miles east of Bishop, and about seventy-five miles north of Death Valley, California. It's the only important—it's all there is—in this valley of about a hundred square miles, except a small road [and] mailing station. There was a normal staff of about four, except there were a lot of visitors. The total student body is about twenty. About ten were introduced to us as freshman on the basis of personal interview or application. There's no cost in going to the school. And about five of these are either unhappy and leave or are asked to leave.

They had trouble getting science teachers because a faculty of four could hardly have much science, and during the '30s, there was nothing scientific about the place in facilities except a small museum. My contact there, my being selected by Pauling dates also back a bit. Many years before, in the early '30s, the inorganic chemist, Neville [Vincent] Sidgwick (1873-1952), as Baker Lecturer at Cornell was invited to stay at Telluride House, supported also by the Telluride Foundation, with money left by L. L. Nunn. He was encouraged on going west to visit Deep Springs. I think that Sidgwick enjoyed his visit to Deep Springs so much that he got [James Bryant] Conant (1893-1978) to visit there. And advice was sought from Conant as to what to do about getting a part-time teacher up. And Conant said, "Get Pauling up, and then get one of the younger post-doctorals from Pauling's lab." So Linus and Ava Helen (1903-1981) Pauling went up there in 1936, and liked the place, and they suggested that I was a post-doc and didn't have regular classes at Caltech and I'd be a good person to teach freshman chemistry to the two students who wanted it. This was the fall of '37.

By this time I had met Grace Mary Seeley and we were hoping to get married when I was out of debt. [They married in Flagstaff, Arizona on December 2, 1937.] I didn't mention my debt. But my European trip in 1933-34 was partly financed by a loan from A. A. Noyes. At this time in my life, with a PhD a year under my belt, I had a debt of thirteen hundred dollars. In my second post-doctoral year I had an income of seventeen hundred dollars, paying thirty-five a month out for the support of [daughter] Patty.

Grace Mary's brother and sister-in-law had come from Colorado to take her back. She had to leave California because of a death in the family, and I went to San Francisco with Grace Mary and her brother and sister-in-law. And then Grace Mary and I parted in San Francisco and I drove on down to Deep Springs. There I found that there were two of the third-year men, Robert Henderson and Robert Sproul. Sproul is now a professor of electronics at Cornell and Henderson is now in chemistry in industry, but he later got a PhD in Chemistry at UCLA.

I was very hospitably received at the school. The dean at the school at the time was a young man, Lawrence Alpheus Kimpton (1910-1977). A man of the staff, Armand W. Kelly (1911-1968), was an economist, but he'd had a couple of years of math and two years of chemistry at Cornell, and he was prepared to carry the burden of the course. Henderson had cleaned the museum out and made a fairly respectable laboratory for four or five men. My job was to set the direction of the course, so I had a total of five or six visits of about three days each. The first two were a little longer. I lectured about four hours each day each visit, and left behind outlines for the lab and the like, and did some work by correspondence. These fellows accomplished a good heavy year-and-a-half course in chemistry.

The lab text I used was one by A. A. Noyes. It had just been dropped at Caltech and is used nowhere. As an experimental course which puts a lot of responsibility on the student, it represents elementary physical chemistry, rather than freshman chemistry. And it's one that helped thrill me with chemistry when I'd been a freshman at Caltech in 1929 and '30.

Kimpton was a young philosophy student from Cornell, who'd been several years at Deep Springs. He'd gone there because during the Depression he hadn't been able to find anything else.

I taught the next year the same sort of course, just part-time, but Kelly had by then had experience and it was satisfactorily handled, to eight students of the then freshman class, the fall of 1938. UCLA didn't like my outside teaching so that I had to agree to drop it after the first year. I went to UCLA as an instructor in 1938. Eight violated the Deep Springs rule for class size, so there were two sections of four each.

I should say a few words about the college. The students paid no tuition. They went three years for two years of college credit, but they worked half-time maintaining the ranch. The student body had all the legal power over the ranch. Nunn was dead, but his brother, Paul N. Nunn, was an old man who wasn't very closely involved in it, but he used to like to fight with the students. Quarrels between the students and faculty were common on matters of responsibility. The students had to learn about range problems, and had to learn to work in the kitchen, the laundry, the garage, and ride herd. They had to stay there except for a short period in the spring. Some of the boys were paid to stay during the summer vacation.

But I liked the idea of putting substantial responsibility on young people. And the only limit of their responsibility was that they had to set up a criticism of the mistakes they made. I also liked treating the students as equals. There was a very argumentative environment up there, but the faculty never pulled rank on them. I liked this very much and I would have liked further contact with the college. I don't know what happened to it during the war. They had draft problems. They took high school kids for a while, and I've not had the chance to go back since. But I have kept contact with the Telluride people, and I know what's happened to some of the boys.

Lawrence Kimpton later went to the University of Nevada, and then to the University of Kansas City. During the war, he was Dean of Letters and Science at Kansas City, and he was very unhappy about not contributing more to the war effort. He was also acting dean of the Law School. Whenever he was in Chicago, he would drop in on us. We felt that the Manhattan Project in Chicago had such poor administration that a man like Lawry would be a big advantage.

We talked to R[ichard] L. Doan, who was the administrative officer at Chicago. Doan was planning to leave for Oak Ridge which was about to be set up. This was the spring and summer of 1943. Doan would have liked to have taken Kimpton. I think I had a lot to do with his coming. Doan got him, but he was administrative officer for Doan only a few weeks when the dean of one of the colleges copped Kimpton for an assistant dean. Kimpton went on to be a dean at Chicago after the war, and then for a little while was a dean and vice president of Stanford, and then was called back to Chicago to be vice president a few months before [Robert

Maynard] Hutchins (1899-1977) resigned. And he succeeded Hutchins as Chancellor.

Armand Kelly was in Washington with the OPA [Office of Price Administration] during the war and used to visit us. And just about the time we went to Oak Ridge, he went to Los Alamos as personnel officer [Assistant Personnel Manager, Project Y]. Armand was also involved in personnel in the Eniwetok [test].

While we were in the Manhattan Project at Chicago, we had precious little university contact, except for the University of Chicago people who were on the Project. In the chemistry department, I had the feeling that they were somewhat jealous of our resources and funds, frightened of our demands for space, dubious of what they knew of the goals of the Project, and did very little to make us welcome, with a few exceptions. We were entertained maybe two or three times by members of the chemistry department.

One of the men who was extremely friendly though, was James Franck, who shortly after joined the Project. I never saw Hutchins when I was at Chicago. I was invited to one function when Hutchins spoke. We did have a nominal relationship with the University of Chicago. We had Chicago paychecks. We were nominally temporary staff members of the University of Chicago.

JBS: What was Hutchins' attitude toward the Manhattan Project?

Coryell: The general story was that Hutchins was rather anti-scientific, and it seemed to us from other universities that the position of the physics and chemistry departments at Chicago was rather poor in 1941 and the late thirties, compared to what it was in other universities. The chemistry department was perhaps worse because an able chairman had died about '36, and he had not been replaced by a leading figure. I think the department was run by committee.

But suddenly, with the announcement of the goal of the Project, Hutchins took a tremendous interest in the implications of atomic energy, and he was one of the early people to make pertinent, and I think very effective statements, about the importance of atomic energy, the importance of its control, and the importance of scientists in having it understood. This began in August of 1945.

I suppose he had some slight knowledge of what was going on in the Project, but certainly not enough knowledge or curiosity to come once in a while into Project circles. And I don't have any knowledge that he had any clearance. Nobody ever talked of having talked over any of their problems with Hutchins.

Now, Kimpton, as dean, who'd had three months on the Project, was involved in decisions involving the relationship of the University of Chicago to the Argonne [known as University of Chicago Met Lab until established as Argonne National Laboratory, 1946], and the University of Chicago to Oak Ridge. [September 1942, chosen site for pilot plants to separate uranium isotopes in Tennessee.] But since I was away in this period, Kimpton himself could tell more about this relationship.

JBS: How much was the Manhattan Project and the two laboratories cloaked in secrecy?

Coryell: This is a problem for the historian to find out. How cloaked in secrecy was it?

It depends on the imagination of the person involved. We had very dear friends next door at Oak Ridge [Coryell home was 137 Outer Drive], the Taylors, Matt [and Lib]. Matt Taylor was personnel director of the Y-12 Project under Tennessee Eastland. He helped hire in a very effective way a large portion of some twenty thousand people that the Y-12 employed. Yet when President [Harry S.] Truman (1884-1972) announced the story of the atomic bomb coming home from the Potsdam Conference, he was as surprised as the newspapers said all of us were. He seemed not to have had the slightest knowledge of what the goal of the Tennessee Eastland Project was.

JBS: How did this security problem affect employment of personnel?

Coryell: Well, the problem did come up of how much someone should say in trying to employ personnel. And I always was very liberal in this with the following background.

As I understand it, when I arrived at Chicago on Sunday, the 5th or 6th of May [probably Sunday, May 3rd], coming from California on the Chief [train], the security rules principle was that, legally, Arthur Holly Compton (1892-1962) was personally responsible for people hired until counter-intelligence had time to clear them which seemed to take several months. This meant, of course, a delegation of authority. The people I hired Compton had never met and probably had never heard of. Obviously I was responsible and responsible to Compton.

Before I left California, I invited a couple of young men to join me on the Project. Edward L. Brady (1919-1987) is one of these. He's presently with the United States Embassy in Vienna, associated with the IAEA, the American mission to the International Atomic Energy Agency. George Campbell, teaching at Pepperdine College presently, and Don[ald William] Engelkemeir (1919-), who is now at the Argonne National Laboratory. I told them very little, except that it was an extremely important project, I think I said involving radioactivity. When I was hired [Frank Harold] Spedding (1902-1984) [selected by A. H. Compton from February 1942 to organize the chemistry division of the Chicago laboratory] said I should read everything in the book.

I was a little uncertain what to do about these things, but Spedding, at Berkeley, had told me quite a lot about the Project and, when I brought in people such as Alexander Langsdorf, Jr. (1912-1996), I went with him to see [Samuel King] Allison (1900-1965). I had brought news to Allison that he wasn't very happy at St. Louis, and he had been approached by Ernest Orlando Lawrence (1901-1958) to go to either Oak Ridge or Berkeley on the electromagnetic project. And I knew enough about that project to know that there was nothing interesting about that project. Our project had all the glory, except that maybe Los Alamos had more, but Los Alamos was still very nebulous. This is early 1943.

But Sam Allison, who had formerly been director of chemistry [at Chicago] and was high up in the physics machine and knew Langsdorf anyway, told him very flatly what we were doing and a fair amount of what Y-12 was doing, and told him what the difficulties and advantages technically of the choice were. Living in Chicago was going to be a bad thing. Fairly technical matters. The same thing was true when I brought Kimpton in to see Allison and R. L. Doan, who was the administrative officer who hired Kimpton. Now, here was a non-technical man being brought in but a fair amount of technical information was being handed over to him. The assumption was that Doan and Allison knew that I knew Kimpton quite well. And they figured I wouldn't ask him to join the Project if things weren't clear. So, when I had to hire people whom I knew, I was pretty frank about a little picture of the overall nature of the Project, so one could see where the fission product came in, because I was hiring people for the fission product elements and the radiochemistry of this was my obligation to the Project.

JBS: How much did your family or friends know about the Project?

Coryell: I was asked by Spedding when I was hired to come and leave my wife away for a month till I got broken in. In addition, we were probably going to move to another site. This was April 1942, March 31 or [Wednesday] April 1. So I left for the Project as soon as I could get away from classes, leaving Los Angeles on [Saturday] May 2 or May 3, 1942. We arranged for Grace Mary to stay with friends at Berkeley and to join me when I knew things were all right, somewhere around the tenth of June.

Mrs. Coryell: I was told to keep a separate address, not to say a single word, and I asked no questions. I said that you [aged 29] were going to Chicago. This all happened with Sam Ruben (1913-1943). He told me then [in Berkeley] that I was a complete idiot to allow you to join up. He said, "You know what those idiots are going to do?" And I said, "No, but anything for the war effort is all right." And then Sam told me they were going to work with radioactivity. I gulped. Sam said they were going to harness atomic power, and therefore I had the idea that there was going to be some terrific explosion. He didn't mention atomic bomb. Sam and Fred and Julie [Stitt] (Fred Stitt, 1911-1997) were there. [JEC is named for J. Stitt.]

Coryell: I understand that during the war and after the Army was most concerned about security at Berkeley. And there's a story that a security officer was sent out to Berkeley to check into this. And in civilian clothing, he went into the lab and walked around and talked to people and asked them what they were doing, and they told him. Then he called everybody together and gave them holy hell for the vast amount of information that they had given to an arbitrary civilian that had come into their lab. And there were stories late in the war about attempted Russian infiltration at Berkeley.

Sometime in early '44, I got a letter from a dear friend at Berkeley, and it ended with a paragraph, "I hope you get lots and lots of U-235." I wasn't interested in U-235, I was interested in Pu-239. But about a few months before the end of the war my mother wrote me a long letter. She had never tried quizzing me, and of course, I didn't see her from the day I caught the train in Pasadena, until after the war. But Mother wrote a letter saying that a minister had given a talk to a church group, and told them the tremendous things that were being done, and the things that were being done to get atomic energy, and that U-235 would be the key to winning the war and would be the key to a great future for mankind. So she said, "Now, Charles, tell me everything you know about U-235." So I wrote back and said, well, it is true, everything she said, that if U-235 could be isolated in quantities, it could be made the basis of a power industry, and probably make a very powerful explosive, but I was working on quite other things. This was really true. My knowledge of U-235 was minimal, except for its relation in the reactor.

About November, 1939, I saw that in *Time* magazine, Hitler had appointed four very eminent scientists to an atomic power panel. I think [Otto] Hahn (1879-1968) and [Werner] Heisenberg (1901-1976) were on this list. A thing that worried me very much during the war was a picture in *Time* magazine, about November, '43, this was before the invasion. There was a picture of a big gun in a cave with drapings to hide it, and the gun was called Urania. We talked about this at Oak Ridge. This was not the sort of thing that we talked about at home because of security restrictions. Several of us felt that Hitler was trying to use the possible drama of atomic energy as a morale booster to Germany and as a threat to the captured people of France and other nations, [so] this name was given it.

I interviewed a young man named James Dial to join my group. This must have been about January of 1943. He was a boy who was very independent-thinking. He was not a very great thinker, but he'd been very excited about atomic energy, typical schoolboy enthusiasm. This was the oddest interview I ever had in my life. This kid knew we were working on atomic energy. He'd guessed it. And mostly it was him quizzing me about atomic energy and how he could get into it. So we hired the person.

It is a legend of the Manhattan Project, you see we had the name, "Metallurgical Laboratory." This laboratory was founded sometime in the early spring of 1942. It was already some months old when I arrived in May. When I first arrived, I'm sure no honest-to-God metallurgists worked for the organization. We hadn't gotten to the point where we had any uranium metal, and we were a long way from having any plutonium metal. But sometimes a metallurgist would apply for a job on the Project. And there were jokes about how embarrassing it was for the personnel officer, Joyce C[lennan] Stearns (1893-1948), and some of the other people, whenever a real metallurgist wanted to work for the Metallurgical Laboratory.

There is also the story of when we first did get any uranium metal, before the shops were set up to handle it, Ed[ward Chester] Creutz (1913-2009), now at General Atomic, took some of this uranium out to do a trick machine job, something they couldn't do for some reason or other in the shops of the Metallurgical Laboratory in Ryerson. And an old German metal worker in the shop, when he saw this brilliant white spark that uranium makes when it's put to a grinding wheel – it looks very much like the Fourth of July sparklers only perhaps the spark is even whiter—he said, "That's uranium metal." The joke is that they had the choice of hiring the man or shooting him so they had to hire him.

JBS: Was there any security restriction concerning the names of plutonium and uranium?

Coryell: Uranium metal and other specialized metals were always given fake names. And uranium, for the people building the first reactors, was always called Molly, a fake on molybdenum [Mo]. It is a fairly dense white metal, and it was thought that people h[ear]ing this who had had some metals experience would not catch on that this was not molybdenum, as they probably would not have seen pure molybdenum before. The high density of uranium, of course, is one of the problems. It has a density substantially higher than lead, 18.9, whereas lead is 11.3.

The code name used by the British project, which started to work on the separation of uranium by gaseous diffusion and was later moved to K-25 in Oak Ridge, was Tuballoy [Tube Alloys]. And this was used by the people who had been on the Project the longest, and the symbol Tu was used. But uranium has an oxide characteristic U_3O_8 [L. S. Goldring], I don't think now of any other metal which has this oxide, and there would be reports of Tu_3O_8 being reduced to TuO_2 by hydrogen. And any chemist would have to decide that this was uranium unless he wanted to postulate it was some metal not yet known to man.

For the chemical code used in [Glenn Theodore] Seaborg's (1912-1999) group, the three successive elements, uranium, neptunium, plutonium, were called copper, silver, and gold. [According to his memoir, Seaborg used copper as a synonym for plutonium. W. Loveland, September 13, 2011.] Copper, silver, and gold are on Group 1B of the periodic system on successive levels, and, of course, are increasingly valuable. Neptunium is only valuable in that it is used to create plutonium, on the way we were working on it. But sometimes we'd be doing a chemical reaction, for instance, uranium metal will dissolve in copper nitrate solution; you'd have code copper dissolving in honest-to-God copper and we'd have to distinguish between code copper and real copper.

Another code which was used which still persists, we would refer to the various isotopes of the various elements by numbers. Uranium-235 has the atomic number 92 and the mass number 235, so we would take the last digit of the atomic number, 2, and the last digit of the mass number, 5, so uranium-235 is 25, or uranium-238 is 28, or plutonium-239 is 49. And already, early in the war, Milton Burton (1902-1985) thought we ought to form an alumni group called the forty-niners, and meet every once in a while after the war.

Going back to security slip problems, I think these may be relatively interesting to historians, and I think they're psychologically very interesting. I was coming from Oak Ridge to Chicago some time probably in late '44 with Lyle Borst (1912-2002). The train stopped in Cincinnati, and Borst had to get off the train to go to the newsstand and buy a copy of one of the science fiction magazines. But it turned out that one of the science fiction magazines had published an article about a secret project, headed by a man named Lizards, which looks like an anagram of [Leo] Szilard (1898-1964), which was working on the Arizona desert on the transmutation of elements to make bombs. The story seems to be that the Army found out about this and proceeded to buy up the magazine from all the newsstands. Borst finally got a copy, but I don't think I ever read the whole article. I think I may have seen it in a hurry and just read the first few paragraphs. But it was obvious that whoever had written the article probably knew a great deal about the Manhattan Project.

I was hired with the understanding that we'd be a short time in Chicago and then we would go to Tennessee to a production site. And when I got to Chicago there was quite a lot of talk about Tennessee. Then, rather abruptly, there was a statement that there'd been a change of plans and we were not going to Tennessee, we were going to Site X. And for some months, it wasn't known to me and seemed not to be known to my boss, Spedding, who let me in on all sorts of stuff, where Site X was. And on Thanksgiving Day, 1942, I had my first chance to see Buck [Brigadier] General Leslie R. Groves, who had the group leaders and division heads of the Project in the conference room in Eckhart Hall at the University of Chicago. Groves was talking about the site qualifications. I think Vannevar Bush (1890-1974) was present and I think Conant was probably there. This was about the time that the DuPont Company was deciding to go in. I think Crawford [Hallock] Greenewalt (1902-1993) was there, and several other DuPont representatives. There'd been a few DuPont people there on loan at Chicago to learn what it was about, but DuPont was making its decision in November of '42.

And Groves said, "By God, we'll buy the whole state of Tennessee, if we have to!" And then he showed where the Clinch River went around like this, and where L&N Railroad [Louisiana & Nashville] crossed. He marked the topography of Oak Ridge, where our plant was going to be, where a couple of the other plants were going to be. So it was from General Leslie R. Groves that I knew that it was Tennessee again. We had been scheduled to go in the fall of '42; the actual moving time was August-September-October, 1943.

JBS: You mentioned at our luncheon the threat that the scientists would be put in the Army.

Coryell: Yes, this was the big argument in the fall of 1942. The suggestion was made that we all should be put in the Army. A. H. Compton seemed to be in favor of it, but some of the younger people felt very strongly against it. I didn't know one way or another. I didn't think that I was going to be very good at saluting and wouldn't know how to act in the Army. But there were arguments that we'd be under better discipline, security problems, and the like. But apparently it was killed above us, and I'm very thankful that we weren't put in uniform. The only advantage I can see is that we would have had GI rights at the end of the war to buy houses.

JBS: How were you recruited into the Project?

Coryell: It's related to how much one tells and one doesn't. Many years ago, this dates it back to '36, I became acquainted with a very dear elderly man, Herbert N[ewby] McCoy (1870 -1944), a retired chemist formerly on the staff at the University of Chicago. He was a specialist in the rare earths. At that time, I knew him as the man who had discovered a way to separate the element europium, element 63, from the remaining rare earths, by reducing it [blank] [from an oxidation number of 3+, similar to the other rare earth elements to 2+] form, where it carries with barium.

41

[Correction by William Walters, January 26, 2010.] McCoy had been very successful at separating europium, using this oxidation state, the most stable of the diadem forms. And as a result, McCoy had nearly a kilogram of pure europium, and nobody else in the world had exploited his method yet and had it. So he visited Caltech, and with various people around Caltech, he did the crystal structure of europium sulphate, showed it was isomorphic with barium sulphate.

I found him a charming man, and I spent several hours trying to argue him into going to the highest oxidation stages of the rare earths. It ought to occur for element 59 [Pr, praseodymium] and element 65 [Tb, terbium, both lanthanoids], the most likely. And I also talked about the possibility of doing these in liquid HF. He decided that I knew a lot of inorganic chemistry, and he also decided that I knew a lot about the rare earths. I found later that he was the man who had done much of the thorium radioactive series as a young man at the University of Chicago before World War One. And during the First World War, he'd helped set up the Lindsay Light and Chemical Company plant, which is a major rare earth producer. He also had married a chemist, Ethel [Mary]Terry (1887-1963), on the University of Chicago staff.

Well, he had known Arthur Holly Compton since Compton first came to the University of Chicago, and he was the senior chemical adviser to the Project. He took a dollar-a-year position. He wasn't in very good health, but he was fascinated by the Manhattan Project, and he knew there were many practicable things like preparing large quantities of uranium oxides and the like. He set up with a young man, William H. Johnson [?]. McCoy and Johnson had a little chemistry laboratory set up in a barn somewhere or a shack set up on the Chicago campus, where they were making uranium nitrate, changing it to U_3O_8, reducing it to U^{-2}, and they were playing around with various methods of getting uranium metal. There was no manufacturing of uranium metal anywhere in the country. According to the *Smyth Report* (1945), the Manhattan Project paid a thousand dollars a pound for the first few hundred pounds they bought. And there were some tough problems. But he also gave Compton advice of whom to acquire from the chemical field.

He came to see me about my birthday, February 21, 1942. He wanted to know if I would take a leave of absence from UCLA for the most important war job imaginable. I wanted to know more about it, because UCLA was a place where I was very, very happy. I'd only seen Chicago once, and I thought it was a terrible place. There were some war projects cooking at UCLA which I was already a consultant to, involving oxygen sources for submarines and aviators, and hydrogen sources for balloons, and some things to do with chemical warfare. McCoy would tell me nothing except that I was uniquely qualified for part of this project, and that I could not imagine any project more important than that. He said that the director of chemistry, Frank A. Spedding, would be in Berkeley in a few weeks and would I go and talk to Spedding. And I said surely, that I would go. Spedding came later than I expected, but I got a couple of letters from Spedding urging me to travel by government expense to Berkeley. I had my first airplane ride by commercial aviation.

I went to Berkeley, the 29th or 30th of March; I remember because the day after I met Spedding was [Wednesday]

43

April Fools Day. Spedding was a busy, rushed, stocky man, often out of breath. He had lots of things to do in a very short time. I've since known him well, and I have a lot of respect for him. But he was a rush boy. I met him in a lab associated with Seaborg. I knew Seaborg slightly. This is Glenn T[heodore] Seaborg, now Chancellor at the University of California at Berkeley. I gathered right away that Seaborg had also accepted a job with Spedding, but he didn't stay while Spedding talked with me.

Spedding rather quickly told me we were working on atomic power to make the biggest sort of bomb we ever knew. We had a fantastic chemical job. All sorts of inorganic chemistry; half the elements of the periodic system were involved in purification problems; a whole lot of elements in nuclear fission (I knew very little about nuclear fission); that we'd be moving to Tennessee soon. He said it was a little hard to see how we were going to do this job, but they wanted men with imagination, courage, and willingness to work very hard, that you might have to go in with shovels and lead pants to shovel the radioactivity out of the way. And I said, "Sure, I'll go." The only question was how soon I could get someone to take my place at UCLA. He said that it was all right with him, but to make it as soon as possible to Chicago. He said he thought the government would pay some of our moving expenses, but he wasn't sure about this, but he'd rather if I'd come without my wife for about a month so that I could get the lay of the land of the Project.

I came back home just ecstatic about going on the Manhattan Project. I had friends at Berkeley. I talked to G[ilbert] N[ewton] Lewis (1875-1946) who was the dean of the Department of Chemistry. I talked to Joel H[enry]

Hildebrand (1881-1983) and Wendell M[itchell] Latimer (1893-1955). I knew Latimer better than Hildebrand and liked him, and I had more in common with him. And they gave me the names of some men who might come down to UCLA, because it was obvious from what Spedding had said that I'd be gone a pretty long time. I'd be gone for the whole period of the war and the war looked like a pretty long war, seen from California in 1942, with Hawaii threatened. The Japanese drive looked tremendous.

Another one of the guys I tried to get was Joseph [William] Kennedy (1916-1957), who was the first man to get a PhD with Seaborg. But he was out of town that weekend, and Seaborg said that he would speak to him. A few weeks later, Seaborg sent me a note saying that Kennedy wasn't interested. After the war, I learned that Kennedy had never heard about the job.

I'm trying to decide how to handle the Seaborg quarrel, and I think that I will be frank but not limitless. The quarrel I had with Seaborg lasted long and was hot and noisy, but I declared peace on him in 1949. We've had friendly relations on business matters all through this period. We had major differences in philosophy and probably competition.

I had known Seaborg slightly. Seaborg was already known to be going to Chicago, but I didn't spend much time with Seaborg. And I felt like an utter neophyte. My knowledge of radioactivity was practically zero. I'd had a text on radioactivity, a German text, in my junior year. [Probably this was S. Meyer and E. Schweidler, *Radioactivitaet*, G. Geugner, Berlin 1927.] I had had one quarter of nuclear physics from Carl D[avid] Anderson (1905-1991) in my first graduate year at Caltech. For some reason or other the Chemistry Department at Caltech wasn't interested though it was pretty active in the Physics Department. I had seen a Geiger counter being operated at Berkeley by [Emiglio] Segrè (1905-1989) and Kennedy while Seaborg was operating the show, and I'd seen Norman Elliott (1908-1985) using an electroscope to measure the thickness of a film for X-ray crystallography. Spedding had told me to read everything in [blank], and I got that book pretty quickly and was going strong on it. [This citation could have been for either E. Rutherford, J. Chadwick, and C. D. Ellis, *Radiations from Radioactive Substances,* Cambridge University Press, 1930, or G. Hevesy and F. A. Paneth, *A Manual of Radioactivity*, Oxford University Press, 1938, from W. Walters, email, January 28, 2010.]

But at any rate, I don't imagine I talked any length with Seaborg, but he knew roughly what Spedding had told me and was counting on me being in his group. He was a sort of super group leader, and I was to have a group under him, the way things were left. Not Seaborg telling me, but Spedding telling me, this becomes rather ticklish later on in the story.

But I knew Saturday, which I think was April 1st, to go visiting various other friends of mine whom I know at Berkeley. I guess the Stitts called and invited me to their house. And Sam Ruben came up. And finally, Ruben said to me, "Charles, I hate to have to tell you this, you're such a nice guy, but you'll have to be careful of Seaborg. He is a perfect son-of-a-bitch." This surprised me, because I thought Seaborg was a very able man. I knew he was a very cold man. But I heard a little bit more about the way he was said to have treated Kennedy and [Arthur C.] Wahl (1917-2006), using a draft situation to browbeat a student who wanted to move to another faculty member. But I decided I would live with the devil if I had to.

I told you the Kennedy story. I'm not sure he would have gone to UCLA, though, because he was prominently involved [in] the projects for Ernest O. Lawrence, as was Martin [David] Kamen (1913-2002), another Seaborg enemy. Kennedy was, of course, later very prominent at Los Alamos and then at Washington University at St. Louis. He died of soft tissue cancer about three years ago. He was one of the finest nuclear chemists this country has produced.

Seaborg produced some superb chemists. Seaborg never tolerated incompetence around him. And Seaborg, I think, knows more nuclear chemistry than anybody living. Fermi pointed this out a long time ago. Seaborg was the best nuclear chemist in the whole Project.

Well, I got away from UCLA as quickly as I could, in early May. I wrote a final examination on the train on the way to Chicago and studied [blank], [the textbook on radioactivity]. And when I arrived at Chicago, I moved into the Miramar Hotel, where I found Seaborg, Spofford G. English (1916-1981), and Isadore Perlman (1915-1991).

Spofford G. English did the most to help me find out what the Project was about. The next day he gave me some reports which are some early reports on the possibilities of production of plutonium. I didn't know for a few days more that it was against absolutely the security rules to take documents home. But I took them home to the hotel, and read them, and put them under my pillow that night. I was tremendously over stimulated by the tremendous new concepts, and by then I was getting still more a feeling of the magnitude of the goals of the Project. The level of excitement went up and I wore myself out by long hours and fast talk and running about, and also by a tremendous guilt complex. I knew so little and I had so much responsibility. I found for instance that I was not a group leader under Seaborg, but that I was co-equal with Seaborg. As I had no men yet and no knowledge of nuclear chemistry, Isadore Perlman and Spofford English were assigned to be liaison with my group.

Now, I had already asked a couple of boys at UCLA if they would hold themselves ready. As I expected, there was a shortage of men, and I wrote letters off to them right away to come. And I also heard that Nathan Sugarman, who was a post-doctoral man with [Josef "Gus"] Fried (1914-2001) at the University of Chicago, would like to join the Project. And I talked with Sugarman within a few days. I had met him in 1941 at Atlantic City at a meeting of the American Chemical Society and had a lot of regard for his abilities. We became extremely close friends, and he was my right hand all through the war years, even when we were separated, when I was at Oak Ridge and he was at Chicago. He eventually went to Los Alamos for a few months and carried out the radiochemical measurements on the efficiency of the Trinity bomb [July 16, 1945].

When I arrived in Chicago, I found that I was to share an office with Seaborg. I have no objection to talking about the Seaborg case so long as I don't stay forever on it.

JBS: It was to a certain extent representative of the quarrels building up from the tensions of a wartime project?

Coryell: There were less fights than I expected from what I knew about human nature then and now. Because I think scientists try to be on the whole objective, and try to be above pettiness. On the other hand, under the tensions, if men decided to fight or were impelled to fight, they fought with extra pressures. And the conflict between Seaborg and me was relatively unimportant to Seaborg. It was the most important thing I could see to maintain the integrity of my group and freedom of action in the assigned framework. I am an emotional person and I threw myself into this extremely heavily.

What happened was that the only way we could have this independence of action was to have action of high quality. For reasons that seem now to me to be more than accidental, I was never able to hire any man who had any radiochemical experience. There weren't very many in the country. But it always happened that Seaborg would suggest somebody, and he'd agree to negotiate, or I'd negotiate, and the man would always end up in Seaborg's group. So that of the twenty or thirty men whom I hired when I was at Chicago, not a single one had had any more experience in nuclear chemistry than I myself had had. But Seaborg would offer to help get them, and either they wouldn't come, or when they came it would turn out that a local crisis would mean that Seaborg got them. I lost Harry [Harrison S.] Brown (1917-1986) to Seaborg. [Edward] Walter Koski (1916-2009) never came. A guy named [Albert Wallace] Hull (1880-1966) never came. Turkevich he almost grabbed from me, although Turkevich had never had any nuclear chemistry.

We learned because we had to learn. It was all right. What we learned we learned well. And starting from nothing, we lifted ourselves by our bootstraps as a group effort. And we had a tremendous esprit de corps among the group.

JBS: What were the particular goals of your part of the Project?

Coryell: When I arrived in Chicago, I don't think I remember what Spedding told me about the goal. I think I was connected with the fission product work. McCoy thought I knew a lot about the rare earths, and these were an important part of the fission products. But I had a fairly general knowledge of inorganic chemistry. Of course, Seaborg had a really superb knowledge of this, and I felt embarrassed to find myself on the same level with Seaborg. It happens that we're about the same age; I am a couple of months older [February 21, April 19, 1912]. We had an identical status in the University. I guess we were made instructors in the same year, certainly assistant professors in the same year, he at Berkeley and I at UCLA So it was easy to equate one with the other.

The personnel officers were fairly sticky about what status you had in the world before you came. At the University of Chicago, there were very few other people with university status in the chemistry group. George Edward Boyd (1911-2006?) was one of these. He was co-equal with Seaborg and me. He was largely in charge of analytical work, though later on in the Project. Lots more analytical work sprung up elsewhere.

He had a deficit in salary compared to us. We had $150 a month extra for away-from-home salaries. We had our university salaries plus a small adjustment for summers. Anybody who was not a resident of Chicago got $150 extra. It was a long time before Boyd overcame these differences.

The fourth man in the group at Chicago who was hired shortly after I came, was Milton Burton, from New York University. But Burton was roughly ten years older than Seaborg, Boyd, and I. He had been in industry and was only an instructor. There was discrimination against him by higher officials in the Project who felt that he was too old and only an instructor. These bright boys are better. But Burton was an awful scrapper. He had been at Berkeley for a year and was a close friend of a lot of the Berkeley people and admired Seaborg very, very greatly. In later squabbles, he would blow hot and cold about decisions.

The director of chemistry was Spedding. But he had had, until January or February, 1942, also no radioactive background. He had a PhD from Berkeley and was one of the winners of the American Chemical Society prizes for bright young men. He was a very hardworking man. He also maintained a big group at Ames, which was his home base, Iowa State College, now University. He had obligations there for uranium metal production, and he did a spectacular job at that. He also played things with a sort of peasant shrewdness, because he didn't feel at home at Chicago with powers like Seaborg whom he couldn't control. Seaborg was under him. He also tried to build up a plutonium group at Ames.

Seaborg was responsible for the chemistry, production chemistry of [radioactive metallic transuranic actinoid elements 93 and 94] neptunium [Np] and plutonium [Pu]. I was responsible for the fission products which are made when you make plutonium. We made our plutonium by loading hundreds of pounds of uranyl nitrate hexahydrate in masonite boxes in the Washington University cyclotron at St. Louis. The cyclotron ran day and night, first under a man named Alexander Langsdorf and then under a man named [Henry] Fulbright (1919-2009), a kind of a hard-driving, Seaborg-type man [who] made Langsdorf unhappy. [Uranyl nitrate $UO_2(NO_3)$, a water-soluble uranium salt, can be prepared by a reaction of uranium salts with nitric acid. The yellow-green crystals of uranium nitrate hexahydrate are triboluminescent. Uranyl nitrate is soluble in water, ethanol, acetone, and ether, but not in benzene, toluene, or chloroform. From W. Walters, email, February 11, 2010.]

JBS: How did compartmentalization affect this work?

Coryell: We had two obligations. The Project had a theory of security, which people at Chicago said was devised by Gregory Breit (1899-1981), a Russian-born scientist now at Yale. He wasn't terribly popular at Chicago, particularly the rules were not. This was all pre-Army. The rules of compartmentalization were already present in the spring of '42. The Army intensified all these things. You were supposed to tell a man only what he needed to know to do a job. You were supposed to tell him all that he needed to know to do a job. And I consistently interpreted the second as more overriding than the first, as long as you trusted a man enough to hire him. All my boys knew everything that I could understand to tell them, that they could cram in their minds, which in any way involve fission products or plutonium. I never knew details of Los Alamos; I never asked questions about timing. The big secrets, it seemed to me, were production quantities and timing.

But everything inside of Chicago that I was not forbidden to tell my group, I would tell my group. We had seminars every Wednesday afternoon and every Saturday afternoon. We worked a six-day week, and many weeks we worked four or five nights. I lost thirty-five pounds in the first six months of the job. At these seminars, we were learning radioactivity, teaching it to ourselves.

Seaborg gave a lecture series every Tuesday night, for a lot of people, a fairly good lecture series. I offered to take the notes, and this way I would learn everything he gave. I would then work that night until about three, and then I would talk to [Edward] Teller (1908-2003) the next day. And I'm sure that the notes are better than his lectures, although under the circumstances, those lectures are very good. This set of notes was used for a long time for training. It was declassified at the end of the war.

Outside of that, we got things out the hard way. It turned out like Deep Springs. You give the young men responsibility and incentive, and there's just no limit. The situation put so many demands on people, that chemists that I hired that I thought from my academic experience would just be middle-road chemists developed sparks of leadership. If they didn't have imagination, they developed solidity and the like. And their relative position in chemistry was greatly elevated in this atmosphere.

Now, part of the conflict of Seaborg and me was native in the personalities. He was extremely competent, had around him extremely competent chosen lieutenants. He told people what he chose to tell them. He got work out of them because they were good men. But he never told people as much. And there were lots of people in the lower capacity who were slower in developing than mine [compared to my group].

Don't forget, there were no more than two dozen nuclear chemists before the war, and Seaborg had probably a hundred and twenty-five by the summer of '43, and by the time that I left Chicago at the end of '43, I probably had about thirty. Part of these moved to Oak Ridge, and we built up so that another forty joined me in Oak Ridge. There was a tremendous educational program. All I can say is that under Seaborg, the educational program went slower. This may be inevitable, because the groups were larger.

But Seaborg also used his position with inside information to delay other people getting knowledge. It seemed to irritate him that there was a separate group at Ames working on the same sort of problems that he was working on. And that Boyd had substantial interest in the field of plutonium chemistry, and indeed, did a lot of work in ion exchange separations, involving this. And I know cases, which I don't believe I should go into now, where there were positive road blocks put in the way of these people, by incomplete information that was passed to them, or by efforts to deny them right to the C-N reports. All the reports at Chicago had 'C' for Chicago, and 'N' stood for 94, the element plutonium. No one group can do a perfect job when so many decisions have to be made on so little evidence. Fundamentally, it was right that the Project have independent workers, as at Ames and as in Boyd's group.

We never got any feedback from Los Alamos. But certainly there was independent work at Los Alamos. All information fed into Los Alamos, and practically none fed back the other way. Except when Los Alamos felt that it was necessary to do their job better, they fed back information so that we'd be better stooges for them. But there was no bitterness about this. We had precious little contact with Los Alamos people.

But when Oak Ridge started up, Seaborg had planned to go to Oak Ridge, and the fight involving me and perhaps some other fights, came to a crisis in the summer of '42, and Seaborg elected not to go. I was very much relieved, and I went. Sugarman took over my group in Chicago when I left. Sugarman didn't have any love for Seaborg, but he didn't see any reason to carry on a divisive fight, and so the Seaborg fight was not carried on at Chicago. I had greatly mistrusted Isadore Perlman, who had been Seaborg's right-hand man at Chicago during the Seaborg fight. But I saw no reason to attack at Oak Ridge, unless I saw something dirty going on. And the people in Seaborg's group had only cordial relations with my group, just an entirely different atmosphere. In time, not long after, there was another group of the same sort out at Hanford. Seaborg also had another group at Berkeley.

The group at Berkeley worked independently of Chicago, and were out of the mainstream of knowledge. Little of their work directly affected Chicago chemistry except in so far as Seaborg developed it at Chicago.

55

Perlman's group at Oak Ridge had somewhat different problems than Seaborg's group at Chicago, because they were close to the reactor, and had first to get these [samples] of plutonium. But all of Perlman's stuff was correlated with Chicago, but was also given an independent check. By late 1944, there was enough stuff to keep busy and the possibility of too much of a one-man show didn't get in the way.

There exists, as a matter of fact, documentation for the most critical months of the Seaborg fight in a secret notebook that I had to leave behind when I left Chicago. Things got so hot that I decided to keep a day-by-day record of the battle. I had been warned at Berkeley that I could expect trouble. But I didn't think very much about it when I moved into Chicago. And I shared an office with Seaborg for a few months. But late in the summer of '42, some awful tough problems came on my section, and I began to find out about the roadblocks put in the way of anybody else who wanted to work on 94. I got fairly friendly with Boyd.

Before the end of the summer of '42, Seaborg had gotten Spedding removed as head of chemistry. Seaborg refused to be responsible to a man who had as little nuclear knowledge as Spedding. Seaborg maintained the right to report directly to [Arthur Holly] Compton. But Spedding remained director of chemistry at Chicago for Boyd, Burton, and Coryell. And Spedding was not resident director at Chicago, because he had the obligation at Ames. It turned out that he spent about half the war in a sleeping car, because he was also promoter for a number of production jobs associated with uranium metal production, the getting of decent calcium, magnesium, all sorts of problems. He came to Chicago about a day or two a week, and he was at Ames a couple of days a week. The rest of the time he was running around.

As acting director, Samuel K. Allison was put in charge. Now, Allison has a PhD in chemistry, but he's an able physicist, and he made no bones about being a chemist. He was, however, an awfully warmhearted guy, and very imaginative. He also had Teller as an adviser. I saw a good deal of Teller, and I have tremendous respect for his imagination and knowledge. I saw Teller about once a week, and he'd outline enough material to keep my group of twelve or fifteen men busy for six months in an hour's time. And the next week there'd be something else. He also helped run my legs off.

Robert [Sanderson] Mulliken (1896-1986), the physicist at Chicago, was appointed special adviser to sort of make peace in chemistry, because some of the higher-ups, A. H. Compton, for instance, began to be worried about these psychoses that seemed to be present. He also visited Berkeley. It was always reported by Seaborg and appeared to be a pocket borough of his.

Allison didn't want to stay on as director of chemistry. He wanted a real chemist in the job. Sometime during the summer Kasimir Fajans (1887-1975) was looked at as a possibility. He is an eminent Polish refugee who came in through Munich. He was the man I would have liked to work for in Munich. I had heard him lecture but never talked to [him], but I got to know him somewhat during the war. I think he was probably not strong enough, rather old for the position. A couple of other people were talked about.

But Latimer was proposed, and I had had an old contact with Latimer and considered him a very able but rather cynical, hard-boiled chemist. I was quite enthusiastic about the idea of Latimer as director of chemistry. But by late summer, I began to feel that things were pretty messy and Latimer was nominally director of Seaborg's group at Berkeley. He also did a lot of gas warfare. He had Sam Ruben and those people working for him. He also was tied in with some project at Northwestern.

Boyd had serious misgivings that Latimer would not be full time at the Project, that Latimer would trust Seaborg completely, and that Latimer's coming would mean that we were all sold out to Seaborg. This must have been about November. Boyd was anti-Latimer. Burton was deliriously pro-Latimer. Seaborg was obviously pro-Latimer. And Allison wanted to know what to do. I decided to vote against Latimer, just because I had no evidence that he would handle the situation in chemistry at Chicago as it ought to be handled, not anything against him as an able chemist.

Shortly thereafter, one of Seaborg's men told me that the penalty I would get if I didn't change my vote back to Latimer was that we'd get another physicist for director, who was known to be dishonest, a man who had the Nobel Prize and then to live up to his glory had to steal his students' work. The man referred to is James Franck. I knew Franck a little bit already, and I knew a lot about his reputation. This angered me, but I heard exactly the same thing from Seaborg himself the next night. Seaborg wanted Latimer badly, and he didn't want another physicist. Of course, there are good reasons for chemists not trusting a physicist as chairman of chemistry. Not that Allison was incompetent, but the problems of chemistry are different from the problems of physics. The Manhattan Project at Chicago never had enough chemical resources for chemists to have full responsibility for chemistry. The physicists were dreaming up ideas faster than we could handle them, and the engineers were setting deadlines and forcing all the issues. The chemists were ground between the millstones of physics and the DuPont Company.

This went on all during the war. This I blame partly on the lack of imagination and confidence of James B. Conant. Because the National Defense Research Council, of which he was the head of chemistry, never released manpower from things like gas warfare to us until late '44 and early '45, we were the last major war project to get going. The problems of physics attracted physicists; whether secrecy occurred or not, the physicists found out about it and flocked. This wasn't true in chemistry. There was no tradition of nuclear chemistry, and there was no easy way for chemists to find out about the Project.

When the DuPont Company moves in, it brings with it nearly eight hundred chemists and practically no physicists. But it was a force for engineering and had a completely different emotional approach to the job. There were many conflicts between the DuPont Company and us. I now feel that these conflicts were inevitable and that they were good. Out of these conflicts we got a better job done. We were the watchdogs for fundamental science to be sure that their processes would work. But when their processes were properly scientifically based, enough manpower and sweat will make them work, and the DuPont Company was wonderful at this.

We never could come to a decision about when is a process good enough to start building a half-a-billion-dollar plant like Hanford. The DuPont forced issues repeatedly along these lines, sometimes prematurely and they had to reverse.

Well, Franck became chairman in early December, '42. I found him wonderful. Franck had his problems with Seaborg. He was stubborn where he had to be, and he gave Seaborg full rein while it involved science.

The biggest issue which I consider related to the problem of the development of the Manhattan Project and Seaborg came up in the choice of the process for the Oak Ridge plant which was to be the pilot for the Hanford plant. Early in the game, probably about the time I arrived, DuPont loaned some m[e]n for chemical technology under a man named Charles M[ilton] Cooper (1900-1971). Isadore Perlman of Seaborg's group was assigned to be liaison man with the engineers. But the DuPont Company didn't take official obligation for the X-10 plant at Oak Ridge and Hanford until November. But I expect that they were well enough committed that no one doubted they were coming in.

According to the knowledge available in the spring and summer of '42, the only way anybody was getting plutonium separated from fission products was the process originally proposed when plutonium was first discovered by Seaborg, Kennedy, Wahl, and Segrè: the use of lanthanum fluoride as a precipitant to carry plutonium in its reduced state, and then the dissolving lanthanum fluoride oxidizing the system. When plutonium goes to the form of plutonyl ion, under which form it has a soluble fluoride and the lanthanum is taken away from it. And this so-called lanthanum fluoride process was more or less officially accepted by the DuPont group, more or less recommended by the Seaborg group, without reference to any other chemical judgment.

The DuPont Company built up quite a core of people. And they were working in tubs, and things like this, and so-called semi-works, working on the hundred-pound level. And they had uniform miserable failure, over many, many months.

The first job Franck had of any scientific nature, that I can recall now, was the problem of what to do about the mainline process for the Oak Ridge plant which was under construction and for the big job at Hanford. Franck insisted that all the chemical judgment of the Chicago Project be brought to bear on this subject, and that no previous commitment was official. The DuPont Company insisted that the chemistry department make an official judgment, as an engineer would, and stand its reputation on it. Franck said, "I am no chemist. I'll give you the best advice of all my chemists."

By then Crawford Greenewalt was in the picture. Greenewalt talked to me and to my senior men, and he invited one of them to talk to the people at DuPont. He did the same for Boyd and Burton. Now collectively among us there was a lot of judgment, but only detailed plutonium knowledge in Boyd's group. Boyd would have liked to have seen an ion-exchange process. But it was effectively removed from possibility by conscious delays, with the argument given, perhaps properly, that this was too radical a process ever to be of any value, that it had to be something conventional like the precipitation process.

Fortunately, about December of '42, Stanley G. Thompson (1912-1976), of Seaborg's group, discovered the bismuth-phosphate process. So in the spring, when the decisions had to be made, there was a big meeting and there were some feelings hurt. The chemistry department decided to make no official recommendation, but make all the facts known to DuPont. In about May, the DuPont Company made an official decision, and it was unanimous, party line from not Moscow but Wilmington, that the big boys wanted bismuth-phosphate, and bismuth-phosphate went.

At about this same time, there was a question about who would go to Oak Ridge. In January of '43, Seaborg approached me via Perlman, insisting that it was intolerable that we have fission product work separated from the element 94 work. Perlman told me that this had been cleared with [Horace van] Norman Hilberry (1899-1986). He was high up in the general advisory staff. [He served as personal aide to Director Arthur Holly Compton. A. Twiss-Brooks.] What the recommendation was, was that I would be a group leader under Seaborg, with part of my group working in the Seaborg line of command. And that part of my group did other things, because we did a broader-front coverage of radiochemical uses in various ways, radiochemistry to study the thickness of uranium, fission products in gasses that did not interest them at all, when they were talking about a heating reactor, things of this type.

Sugarman just about blew his top. Sugarman and I went to see Hilberry, who said that he'd never been approached on this subject. We went to see [Eugene P.] Wigner (1902-1995); we couldn't find Teller. A few days later, I saw Teller, and he said, "Wigner has told me the story about Seaborg. It's nothing I don't know, but it's added some local color to it."

Under pressure from Sugarman and William Rubinson, I decided to stand my ground and fight it out, and to maintain an independent group at Chicago, if necessary, and that we would be an independent group at Oak Ridge. Then it was realized that the Oak Ridge labs weren't designed large enough. Seaborg, by claiming that he was going to move forty men down there, virtually had everything in Oak Ridge possible.

In a sense, though, the bismuth-phosphate decision earlier had broken Seaborg's pipeline to Cooper [prepared blueprints for Oak Ridge] and the engineers. Milton Burton was called in the breach to design a greatly expanded building, a research building perhaps quadrupled in size. Thirty-six thousand square feet was finally put up. Seaborg then announced somewhat later that he wasn't going, and everyone was very much relieved.

A few people moved down in August, 1943. This involved Perlman and English, the heavy element people, [Edward Martin] Shapiro (1916-2003) and C. Vernon Cannon (1916-2001), and some of Burton's radiation men. But Burton didn't move and Burton's men were nominally under me. I'm not a radiation chemist, and they essentially had freedom of action down there. They had a fight with Burton when he came, and they had free phone contact with him.

But I defended them and would ask for the raises when I had a chance. I mean I took care of administrative problems of the group. Seaborg came down every month or month and a half. I also went to Chicago about every month, because there would be Project council meetings, and I would have to report from my group at Oak Ridge. [Cf. Rhodes 1985, p. 411.]

But there were no residuals of the Seaborg fight in the Oak Ridge area. And my path didn't cross Seaborg's thereafter very much. I saw him occasionally. There was a Seaborg – [Jerrold Reinach] Zacharias (1905-1986) fight which gave me great delight. To my knowledge, he never ran me down in public. To my knowledge, I never missed an opportunity to run him down in the Project public for a period of about fifteen months. Beginning about December of '42 until the end of the war, I was bitter about the power struggle and the contrast in policies, and the mistakes involving apparent commitments by chemists to a process which was not any good, not properly researched, could not have been better researched in the time, and should not have been committed. I think that these were probably administrative judgment mistakes in the chemists and the early DuPont people.

Seaborg doesn't like to fight in public. He's a man who thinks before he speaks, and he doesn't like hot things. Once I poured for two hours on him, not invective, but all the reasons why a whole lot of people were mad at him, naming names and places. This was about January of '43. Also, he had two or three of his men, including Bertrand Goldschmidt (1913-2002) and Perlman, in August '42, do a scoop job of all the areas we had responsibility in, and make a major report in the field, a lot of cream skimming. We weren't informed of it until the day it came out in mimeograph. And that day, Seaborg conveniently left for a two weeks' vacation in Oak Ridge, to Tennessee to look over the land. At that time Oak Ridge wasn't there, but he knew it was [to be] there. At that time, I didn't know that Tennessee was back in the works. This Perlman-Goldschmidt report offended a lot of people, and Seaborg was penalized by having to admit me to his Project council meeting, every Thursday night.

I attended. I always had to have something new to present on our side, because there were awfully good men down there. My guys worked like mad while my neck was out a mile. And every week we'd get something new to say that was worth saying. I would occasionally make a nuisance of myself.

I declared peace on Seaborg in 1949, September. Then I was mad right afterwards. We were having dinner together in Atlantic City at the time of the meeting of the American Chemical Society, and we were guests of McGraw Hill, because we were both authors of a book in the National Nuclear Energy series. And I found out a week later that Seaborg was on his way to Washington. He knew that the Russians had fired off a bomb [August 29, Joe-1, Truman announced it on September 23, 1949], and he hadn't given me any advance notice of it.

He did cheat in 1942-43, but he more often cheated by half-information than by misinformation. This was to protect his position which he considered indispensible to the war effort. And we at Chicago who were against him, and the Berkeley people who joined forces with us in this winter and spring, felt that he wanted to go on and dominate Hanford. He had Chicago in his pocket, chemically. He was going to Oak Ridge, which we stopped. He would go to Hanford. It turned out he didn't want to try to go. The fight sort of petered out. We all thought he'd go to Los Alamos. But we were informed that the guards at Los Alamos had been warned that they were to fire if they saw the whites of his eyes!

He visited Los Alamos right after my big blow, late January, early February of '43. But we found out only at the end of the war that he'd gone to Santa Fe, phoned Oppenheimer, and Oppenheimer had come down to see him. He never went inside the gates of Los Alamos until after the war was over. Because Kennedy and Wahl and Segrè had feelings about Seaborg probably akin to my own.

JBS: Do you recall any difficulties in going ahead with the research because of the original six-month appropriation of funds running out in June, 1942?

Coryell: The decision to go ahead on a number of fronts in atomic energy was made on December 6, 1941, in Washington, D. C., the day before the attack on Pearl Harbor. General [George Catlett] Marshall (1880-1959) was one of those attending. I don't know how many government officials were there. But I've often said that maybe some of the confusion on December 7, in Washington, was due to the fact that some of the officials in Washington were just dizzy because of what the scientists had been talking about.

The decision was made to make a six-month effort without any limit on funds. I was only slightly embarrassed when I arrived at Chicago in May. I met Compton after a few days, as soon as he was in town, and I found out that the Project had only full authorization to go until six months after December 7, which was June 7. But nobody had any doubts that it would go on further. But I was one of those who said that if we had to, I'd go back to $75 a month. I wasn't going to leave this Project, just because its funds got cut back a little bit.

The general features of the Manhattan Project are roughly as follows. At Columbia University there were large laboratories which had the name SAM [Special Alloy Materials], which were doing research in so-called gaseous diffusion project. Uranium hexafluoride would pass through thousands of membranes being pumped around, so the heavier form U-238, with six fluorine atoms, was slowly dragged behind the lighter, and you'd eventually get pure U-235. This is a fantastically difficult job. It's easy conceptually and it had been planned in England, and the British plans were moved to the United States. The gaseous diffusion process is the surest way to get U-235, and at present it is the most economical way. But it was a tough job to research.

Harold C[layton] Urey (1893-1981) was in charge of the scientific side of this program. The SAM lab was just starting. The production work was assigned to the Oak Ridge area, to be away from the seacoast and easy to defend, and in an area where there would be plenty of power. This process takes a tremendous amount of power. I'd guess that at the end of the war, that probably, with the wartime inefficiency, probably each kilogram of U-235 made by either the gaseous diffusion or the electromagnetic process, must

67

have cost about as much power as they would make, nearly a million kilowatt days. And the fission of a kilogram of uranium a day gives a million kilowatts. Of power consumption, I'd say about half the power consumption of Boston.

So power requirements set where the project should go, and the British were glad to move this project to the United States, because of its isolation from the war front, and because of these tremendous power requirements, and because the manpower requirements were also so fantastic, and they just didn't have the manpower to do this sort of thing. They had to do the bread-and-butter affairs of radar being developed and mine detection, all sorts of other ad hoc jobs for the midst of battle.

But there was good British participation. This good British participation included Klaus Fuchs (1911-1988), whom I never met but whom the chemists all hated.

The electromagnetic separation of U-235 was an idea growing out of Berkeley and was researched at Berkeley under E. O. Lawrence and scheduled for Oak Ridge, also for power and isolation. The Tennessee Eastman Corporation did the engineering. I think it was ([M. W.)] Kellogg [Company] that did the engineering for the (K-25) gaseous diffusion project.

The physics of reactors had been started at Columbia by Fermi, Leo Szilard (1898-1964), and Wigner. But these people all moved to Chicago because of the crowding in New York City and because of the danger of enemy attack on New York. It was set up in Chicago under Arthur Holly Compton.

According to the stories, the bomb was to have been also sponsored at Chicago, because a reactor is slow burning uranium, which makes plutonium as a by-product. And a bomb is fast burning uranium or plutonium. But before things had gone very far, a separate bomb group was set up under J. Robert Oppenheimer, whose headquarters were in Berkeley until Los Alamos was founded [April 1943]. But Los Alamos was clothed in utter secrecy.

JBS: Was there much contact between the other projects?

Coryell: All these projects were separated one from the other. We had practically no contacts with the uranium projects at Chicago, except as the production of U-235 by gaseous diffusion began to be started in Oak Ridge and the research in New York, a few people began to move to Chicago from this group to do jobs related to our project more than to the gaseous diffusion project. They were scientists whose jobs were no longer necessary as they approached production. I don't remember the names of many of these people. Clarence [Francis] Hiskey (1912-1998) was one of these. He had a messy security record later on. I knew him moderately well, and I saw him some after the war.

Somewhat later, the Naval Research Laboratory in Washington, D. C. developed the idea of what is called thermal diffusion separation, a process requiring fantastic amounts of power. It counts on the very compressed gaseous uranium hexafluoride held between a hot wire and a cold wall, the molecules will go back and forth, and the lighter molecules will rise faster than the heavier. This method of isotope separation was engineered by The

Fercleve Corporation in Oak Ridge [Philip Hauge Abelson, (1913-2004), consultant, for S-50 project]. This process was stopped the minute that news was received that the Japanese had accepted the American request for surrender, it was so expensive. Y-12, the electromagnetic process played a very important part in the providing of material for the Hiroshima bomb. But it was also stopped late in August or early September. The gaseous diffusion plant had no end of headaches.

All these big projects had fantastic problems with corrosion and isolation of the product from crud and gunk. These were good chemical words on the Project. There was a lack of chemical imagination on the Project. They were all physics-dominated except for Urey, and Urey didn't have full chemical say, either, because it became engineering.

Also I think that Urey was somewhat squeezed out, as near as I can see probably in late '44 in this "gray list" story. Because the Army used this term, sometimes other than jokingly, "prematurely anti-fascist." A man who'd been too active in the Spanish Civil War, because in that context in 1935, he was obviously left wing. He might again be too friendly to the Russians in 1943-44.

But Urey was the only Project leader who was a chemist, yet the chemical problems at K-25 were fantastic. Also headaches in inferior production, lubricants, things like that.

In 1942, we at Chicago got a number of British reports, and occasionally a visitor from England would appear. But the only foreigner working with us was a Bertrand Goldschmidt, a Frenchman who had worked with Madame

[Marie] Curie (1867-1934), and he was a very competent nuclear chemist. He couldn't work at Chicago, but he could be loaned to Chicago. He worked about six or eight weeks with Seaborg and then went on to Montreal. The Anglo-French-Canadian project was there. But from '43 on there was little interchange of information with the British. We got little of their information and they got none of ours. There were practically no visits. General Groves seems to have been very anti-British. In spite of agreements made by [Winston] Churchill (1874-1965) and Roosevelt at Quebec, there was just no contact.

I heard stories after the war from Canadians about how they felt. They were promised things and they never got them.

The planning for chemistry at Chicago happened in late January or early February 1942. This was before I was contacted by McCoy. But there is a famous project report, C-C-111, which was written by John Archibald Wheeler (1911-2008), the Princeton physicist who was a very prominent person at Chicago. The meeting involved physicists and chemists, chemical problems and how they relate to physics. This report we used a lot for training purposes, pointing out what should be done, what could be done and what couldn't be done to make massive amounts of plutonium.

JBS: What were your impressions of John Wheeler?

Coryell: John A. Wheeler was extremely active in the Project and a wonderful teacher. Whenever I got a chance I'd spend time with him and he taught me a lot. Other prominent people—Katharine Way (1903-1996), a young woman, was one of the theoretical physicists who also was with us in chemistry a good deal. Teller was in and out of chemistry and physics, and was very stimulating, and I have a tremendous affection for him as I knew him then, and a very high respect for his imagination and abilities.

I can't resist telling a little story on him. He and Fermi were very close together and worked together a lot. But the trouble is that Teller would never stay on one thing for very long. And apparently, Fermi was the only one on the Project who could say, "Edward, stay with this problem!"

Eugene Wigner was involved in physics. He has a very modest manner, and a very sharp conscience. He has a BS in chemistry, but he is a physicist, an able theoretical physicist. And often Wigner would tell you, "I am no chemist, but..." and then he'd tell you something you should know.

Fermi was never very interested in the details of chemistry, but he was always accessible to us. Compton was reasonably accessible, though not very much at the technical level, because he was involved in policy and the like. But in times of tension, as with the DuPont Company, people always said that Compton was a Boy Scout. He always tried to make everybody feel good, and always acted according to the directions of the last man to get to his ear. We really loved and respected him, except that we worried about his being chewed up by the Army and the DuPont Company.

There was a very good esprit de corps among us. We didn't feel part of the University, but one made friendships very fast. Somebody came in and you took him around and showed a bit of the Project. You rarely went out on the town, you had so little time. But the wives got together. And you were close as a brother to someone in a few hours. You rapidly found points of common contact, and you had conferences to help one another. We had a few big Project parties, but not very many of these. But I also noticed a kind of Project snobbery. I, who came in early May, still look a little bit down on latecomers who came in June and July, 1942. I can't quite forget that Burton came a couple of weeks after me, and therefore I outrank him in knowledge.

Burton was one of the most ferocious hirers on the Project. One man, Tom Davis, came through Chicago on his honeymoon. He was fool enough to phone Burton up, and Burton hired him and gave him two more days, and then he came back and started work. Another one of Burton's men got married, and Burton wouldn't give him any time off except Saturday. So the guy got married on Friday evening, came back on Monday morning. But we drove ourselves equally.

Most of the people from my level up belonged to the Quadrangle Club. You'd go to the Quadrangle Club and find people visiting, Oppenheimer and so on. Milton Burton knew more people. He'd travelled around the country a lot; he was older; he had come from the East. And he used to tell Grace Mary and myself, "There goes the most famous photochemist, or biochemist, or this and that and the other thing."

There were lots of biochemists who joined the Project. With the microchemistry of plutonium you need the hand techniques of people who have been getting tissue culture or doing microsurgery. So men like Edward R. Tompkins and Paul Tompkins both came in to join a group with Waldo [E.]

Cohn (1910-1999) whose obligation was fission products in biology, and radiation chemistry. And Paul L. Kirk (1902-1970) was an eminent biologist or biochemist who joined Seaborg's group in microchemistry of plutonium. [Burris Bell] Cunningham (1912-1971) was another one of these.

My major men were Sugarman, more than my right hand, and Norman E. Elliott, a young man whom I'd known at Caltech, from Pomona College. [William] Buck Rubinson. Just before I left Chicago, [T.] Harrison Davies (1912-1997) [same birthday, February 21, as CDC] came. He is also a biochemist, and David Hume (1918-1998), an analytical chemist from UCLA, now a professor at MIT. [Davies' spouse, Esther Farney-Davies was Grace Mary's physician in Chicago and namesake for Julie Esther.]

Seaborg's senior men were Isadore Perlman, his right hand, although Seaborg is about six feet four inches and Perlman is about five feet two inches, a real Mutt and Jeff combination. Also, Spofford English, who is now in the Atomic Energy Commission (1946-74) in Washington, D. C., and who was the closest to me of any of Seaborg's group. John E. Willard from Wisconsin. And a young man who became fairly weighty in the place was Truman P. Kohman (1906-2010), who is now at Carnegie Tech [Carnegie Mellon University]. And one of the most vigorous of the Seaborg men, who is now at Caltech, was Harrison S. Brown. He was very active at the end of the war with Americans United for World Government. And he's been concerned with problems of population control, and oceanography, and Democratic Science Advisory Committee, things like this [including Executive Committee, Association of Oak Ridge Scientists, *Bulletin of Atomic Scientists*, member, National Academy of

Sciences, from 1955, Pugwash Conferences from Third in 1958 plus sixteen more].

There was an extremely democratic atmosphere. Anybody could see Fermi if he could find him in his office. He was always accessible if he had time. He always went to lunch with his coterie around him, which included Herb[ert] Anderson (1914-1988), Leona [Woods] Marshall [Libby] (1919-1986) and her [then] husband, John Marshall, and Gail Young. The Wigner group and the Fermi group, the theoretical groups, were always close together.

Szilard was always everywhere, always knowing what was going on, making friendly, acid comments about things. He always had this curious way of looking at things. He would ask simple questions which were shockingly revealing.

You had in science here, probably more so than at Los Alamos, as though you had a good share of the major painters of the Renaissance all working in one church and all good friends. The fights were relatively rare. Any time you wanted something, someone else had a vacuum pump that wasn't really tied up, you'd get it.

I also worried about the expenditure of money. I had the right to spend up to five hundred dollars on any object. Money was by no means the object. It was time. The first few weeks I'd come home and tell Grace Mary that I had spent so many hundreds of dollars that day, or maybe a thousand. But Sugarman would say, "Nuts, Charles! Isn't the fission product group as good as a destroyer? What do you suppose a destroyer costs?" He'd say, "I'm sure a destroyer costs twenty-five million dollars. I'm going to order eight vacuum pumps because we'll have a devil of a time getting them." I only knew of an expected need of two, but sure enough, the eight would come, and they'd all be used up before we'd get another order.

There were no phone numbers. But there were two or three phone girls, and they knew people's voices. You'd call up and ask to speak to Allison. She'd say, "I'll get him for you." She would know where Sam Allison had last been phoned or phoned. Yet this went on till there were eight or nine hundred people on the Project. She'd locate people just by knowing where they most likely were and knowing where she'd last heard them called.

The Project was supposed to be ultrasecret. It was easy to close off the floor we were on; we had the whole top floor of Jones. But Eckhart and Ryerson were the two physics buildings at Chicago. Ryerson was the old gothic one, Eckhart being also gothic but not so old. Eckhart had the library. The second floor of the library had to be accessible to the people who weren't on the Project. And of course, we had to have a library of our own for the classified stuff. So they built up the center stairs a passageway for us, and a separate door at the bottom. Some parts of the building were walled off from others. And this worried Szilard very much. He was afraid that somebody would get lost in the passage ways connecting various secret parts with the others. He had heard that if you were ever lost in the mountains, if you could find a porcupine and kill it, you could have food. So he proposed that the Army put some porcupines around there in case somebody got lost.

Fermi was sure that he could break into the building without the guards knowing about it. I don't think that he ever carried this out. But he showed two or three obvious ways if a person had any skill in climbing gothic fixtures outside, tricks that an English student can always do, could get in.

A little bit later, General Groves gave a security talk to all the senior men on the Project. He said we had to protect ourselves against a dedicated enemy who would sacrifice himself to achieve his goal. This is what you have to do in security. This is the way that the MIT reactor has to be designed, so that a man who wants to commit suicide and destroy the reactor can still not do serious damage. He impressed me quite a bit. He also reminded me somewhat of my younger brother [William Harlan Coryell, Jr. (1914-1963)] in temperament and build. So after the lecture I spoke to him. I told him that I thought he'd given a very good talk, but what would he think if a man threw a hand grenade through the window of the conference room we were in. In the room had been people like Compton, Conant, Szilard, all the people I've been mentioning. And he said, "It'd be a damn good thing. There's too much hot air in here!" It made me awfully mad.

Session Three — 7 August 1960

Coryell: Today is within one day of being the 15th anniversary of the Hiroshima bomb and like you, I felt it should have had more attention in churches and elsewhere.

I thought I'd talk today about some alarms and rebellions in the Project, particularly ones during the war itself, and before I do this, I think it would be a good idea to give a character sketch of several of the people involved, particularly James Franck.

James Franck was one of the best-loved of the physicists produced in the golden age of science in Germany. He was director of the *Physicalische Institut* in Göttingen for a very long time and he was teacher to most of the brilliant students in physics—that is, men who weren't his own students but would come to Göttingen to work with him in the Franck school. His name is marked with some of the interesting effects in physics with some of the junior authors who later were associated with the Manhattan Project. For instance, Eugene Rabinowitch (1901-1973), a physical chemist who has been heavily involved since the mid-thirties in the physical chemistry and chemical physics of photosynthesis, worked in photochemistry with Franck and the two invented the Franck-Rabinowich cage hypothesis about the way excited molecules bounce around within a cage of cold molecules for a long time before they move to some other part of the solution.

There's also the Franck-Condon principle (1926), which relates to the way a molecule fails to change its internuclear separation on being raised to an excited state. [Edward U. Condon (1902-1974)].

There's a third one of these Franck principles that I think also involves a Manhattan Project personality. I might also say that I didn't know Condon during the war in the Manhattan Project but I could see his position in the Manhattan Project from his testimony before the House Un-American Activities Committee when Congressman [Richard Bernard] Vail (1895-1955) held hearings in Chicago in 1948. He left Los Alamos in disgust at the security situation according to his statement to Congress and I don't know that he had later contact. He was, however, Director of the Bureau of Standards at the end of the war and was the highest-placed scientist in the U. S. government in the critical period of 1945 to 1947, and had lots of influence on the relations of science to government, furnished channels of contact, and was strong against the May-Johnson bill [October 1945, Senator Edwin C. Johnson of Colorado introduced S. 1463 and Representative Andrew J. May of Kentucky, H. R. 4280] and strong for the McMahon bill. [Brien McMahon (1903-1952) of Connecticut introduced S. 1717, December 1945, regarding civilian control of atomic energy.]

Now, Franck was liberal in every sense of the word at all times. I'd like to say that he was greatest man that I know of that I've ever seen and I'm furthermore proud of the fact that I worked with him and that he was my boss for some months—for about nine months at Chicago—on the Manhattan Project. What I admire about him is that he would enter any battle on the side of the right, irrespective of the prospect of victory or irrespective of the state of his health or the strength of his support, and he is totally honest in explaining to the rest of the world where he stands.

Before Hitler came to power in came to power in Germany he had considerable concern about the evil of the Nazi philosophy and its methods and was quite outspoken in this regard. I understand that there was a movement sponsored by a man named Richter in the Ministry of Education in either Prussia or Germany—I'm not sure which, because I think Göttingen is in Prussia, I may say also I've never seen Göttingen. Richter is an Aryan. His daughter, Valborg, married George Boyd [1942], who was my colleague at Chicago and at Oak Ridge. After the war, Professor Richter was called back to be Rector at the University of Bonn—I believe he's still living as Rector Emeritus of Bonn—and he was a very close friend of James Franck. Richter proposed that if Hitler should come to power all university professors of conscience and courage, irrespective of racial origin, should resign. Hitler came to power and nobody resigned except Richter and some Jews. Franck was not asked to resign, but very shortly after Hitler came to power Franck resigned in a public letter to the German government, which I think was published in the Göttingen newspaper and all over the country, saying he could not be a public official, as university professors were, for a government with the philosophy that it had. I asked Franck about this in '43, when we saw each other so often, and he said, "Coryell" (he

81

never called me Charles), "I didn't know what an idiot I was. It wasn't my concern for myself or for my immediate family. I didn't realize what the anger of that government might do to my friends who were not Jewish and my students who were not Jewish." However, it happened that the position of Franck was so unassailable that nothing happened. He stayed with dignity in Germany for some months and retired, with his wife and some resources, to America.

I first met him when he visited UCLA as Hitchcock Lecturer, a lectureship lasting some weeks at Berkeley and then the Hitchcock Lecturer comes to UCLA for a couple of days. He gave several lectures on photosynthesis. When he arrived in Los Angeles I was one of his [escorts]—I took him around and showed him the university and all that, I may have met him at the train with some other people—a man named [Olenus L.] Sponsler (1879-1953), who was head of the botany department but interested in joint physical biology efforts. At UCLA there was a very friendly collaboration between biologists, physicists, and chemists that I was happy about and I was on the guiding committee of this physical biology setup. Franck had a serious case of enteritis and felt terribly, but courageously went ahead with two beautiful lectures on the physical chemistry of photosynthesis. As a scientific aside, I might also add that he was quite bitter about the foremost man in experimental photosynthesis, [Emil] Warburg (1846-1931). Twenty years earlier Warburg had shown that, in ideal circumstances, four atoms of light, four photons or four quanta were the minimum required to change a molecule of carbon dioxide to an atom of carbon in the framework of the sugar which is the main product that the plant makes. This had always posed an interesting problem since it represents a very highly efficient process chemical-wise as well as biology-wise. Only in the very late '30s was there evidence that this was probably not right. Some of this evidence came from Farrington Daniels, at the University of Wisconsin, and Daniels was later director of chemistry at the Manhattan Project. Probably his friendship with Franck helped make him the choice for this purpose. Daniels and some younger collaborators showed that in a typical circumstance near optimum, the number was 10 or 12. Franck was bitter that a finding such as Warburg's finding of the early '20s had stood unchallenged and unverified for 20 years. He also pointed out to me that this was widespread in science. For instance, the positron discovered by Carl [David] Anderson in 1932, I think it was [yes], could easily have been discovered in 1912 when people first had appreciable points of radiance since the gamma rays from radium create positrons. The effects which the positrons made in this instance were an odd sort of background that for simplicity were always subtracted, and people overlooked this. I received this principle from Franck: "Never leave unexplained an odd bump on a curve."

I think I said in my first discussion with you that I have tried to make my own scientific position, personality, and attitudes somewhat of an average of A. A. Noyes, whom I discussed in my first lecture [Session One], Pauling, whom I think I discussed in my first lecture and will later on when we get to the postwar period, and Franck. Of all the men I knew in the Manhattan Project, Franck is the greatest. I failed to see Franklin D. Roosevelt—for some reasons I might think Roosevelt is greater—though I think not. Roosevelt was due to visit Oak Ridge the week he died. [Nichols 1987 rejects this impression.] But I think knowing what I think I know about Roosevelt and knowing what I do know about Franck, I'd put Franck first. Franck was totally honest and Franklin D. was not.

Franck was very concerned about the fate of liberal scholars of all sorts—not only Jews—in the world and he and his wife [née Ingrid Josephson, m. 1911] used up most of their resources helping move friends and relatives out of Hitler's domain. Franck was invited to join up with Chaim Weizmann (1874-1952), who was putting the full power of the Zionist movement into saving Jewish people and offered a haven in Israel for everyone that would come. In Weizmann's autobiography, *Trial and Error* (1949), Weizmann indicates that Mrs. Franck was very much opposed to this, and I know that when I talked to Franck after I returned from Israel [1954] he was very bitter at the distortion of Mrs. Franck's character and social attitudes stated in the Weizmann book.

I don't know now why Franck didn't go there. He was given a berth in the University of Chicago and late in the '30s he was offered the chairmanship of the department. In our many hours together he told me lots of these stories. I'd like to pull back into why he did things. He said, "Coryell, I told them I wouldn't be chairman of the department. I'm not only a Jew, but a damn foreigner. How could I write letters for little Johnnie to work in the steel mills? American departments should be run by Americans."

When the first Mrs. Franck died, which I believe happened in 1941 [~1940], Franck had to sell his library to pay the expenses of the illness and funeral. He was living in the Quadrangle Club, his heart was not good. He was a slow-moving, very quiet man, but of course he had a constant stream of friends and students coming to see him. He had ample money for his research at the University of Chicago under the Josiah Macy, Jr. Foundation and his lieutenant at the laboratory there was Hans Gaffron (1902-1979). And there was a very nice research group.

When I first came to Chicago I had known but three people on the staff personally and I called on these rather early. Franck was one of these and he was quite interested because he remembered our discussions at UCLA. I saw him a few times about photosynthesis.

I spent some time with George Willard Wheland (1907-1976?) [ATB: joined U. C. Chemistry Department in 1937], whom I'd got acquainted with at Caltech when he was post-doctoral. Wheland's wife was the sister of my best friend in high school, a young man who died early. Harold Webster Clayton was his name. He died before he went through college. And I had known Nathan Sugarman briefly from a meeting of the American Chemical Society in Atlantic City in 1941. I also formed a good friendship with Barron [Eleazar Sebastian Guzmán-Barrón (1899-1957)], a Peruvian Indian who was very prominent—it's possible that I had met him before, but he wasn't a member of the Chemistry Department. [A biochemist, Dr. Barron studied health effects of radiation on the Project, discovered the cytochrome enzyme, worked on effects of high altitude on metabolism, nutrition and cancer research. He died of multiple myeloma. ATB email from Isabel Guzman-Barron, his granddaughter, November 18, 2011.] We were treated rather warmly by Warren C. Johnson (1901-) and later on he became my boss, but most of this period I was at Chicago Johnson was away.

There was an organic chemist, head of the *Journal of Organic Chemistry*, Otto Reinmuth (1900-1956), who was nice to us. And T[homas] F[raser] Young (1897-1977), who was later in the Project, was nice to us. But the rest of the department was essentially cold or hostile.

I was a guest at the department seminar once or twice—I think I spoke on generalized acidity with demonstration, but I'm not sure right now—and I gave a seminar in enzyme dynamics kinetics to Barron's group, but I had very little feeling of being in a university because of the tremendous pressures on us and the way the project was introverted to its own problems.

I now have left Franck a little bit in one of my digressions. I'll just say a little more about this atmosphere in the early days of the Project. I've forgotten now what I was going to say.

JBS: You said something last time about the problem of the separation of the group of the Manhattan Project from the University of Chicago.

Coryell: We were unwelcome guests who threatened their space and threatened their glory and had too much money and were obviously wasting our time. We had our own internal education process—I don't know if I mentioned that Seaborg ran [the group to "separate plutonium in the amounts and purity required for war purposes." Smyth 1945, p. 100]—I did last time? All right.

There was a period in the Chicago days when I had commitments five nights of the week, and of course

Saturday was a full working day. During the early days of the Project I lost 35 pounds. Though I loved what I was doing, I was just scared to death at the obligations and felt that every shortcoming of my own was a potential delay in the war and that every failure of our group to do the brilliant things that were outlined by men like Teller and Mulliken and Allison were just a measure of our inadequacy in the most important job in the war and that each delay and each loss could be measured in tens of thousands of lives and tens of millions of dollars. I think I mentioned also how Sugarman cheered me up on spending money. A destroyer costs much more than we could spend.

Well, I'll go back. Franck was marked to become chairman of chemistry largely because of my failure to support Latimer, and I've remarked that some of the opposition—which was really a natural opposition, I think, to getting an unknown man who was not only a physicist [but not a chemist]. (All of us chemists resented being bossed by physicists because we felt that physicists were wonderful people and they did beautiful work but they always underestimated the chemical nature of the problem or chemical difficulties). And the Manhattan Project in every one of its arms had serious roadblocks because of lack of chemical imagination and chemical understanding on the part of both the physicists and later on the engineers—I think I'm repeating also the last discussion.

Franck, I believe, had trouble being cleared. The clearance was done, in the last analysis, by counterintelligence in the Army and it was extremely difficult for an army officer to realize that a man who had been a captain in the Prussian army would now serve on our side. I may say that Franck was contemptuous of army operations and army administration, but he said there was one good thing about the matter. He knew both armies and Germany had a worse one.

I didn't tell any army stories about Franck. Franck was inducted into the Kaiser's army in 1914 as a private and every time he went up in rank he was more unhappy. He was happiest as a private in the German army. He said, "The young man doesn't fear death. He has lots of time to think about physics." But he finally came to the rank of captain, and whenever he was in trouble with the colonel the colonel would call him "Herr Professor." He hated the way the colonel would explain things to the troops and the colonel was very upset because Franck got no Iron Crosses for his men. So Franck finally checked with other captains and found out that you had to swear that the man had done everything possible for God, emperor, and country, and pretty soon he began getting more Iron Crosses for his men than anyone else.

He also had an interest in the human side of his troops. There was a man named Hinfpater, Franck told me about once, an old sergeant who kept the company funds with which they had USO activity—I mean sort of pleasures on leave periods. Hinfpater was so stingy with the money that the men were grumbling and Franck tried to force him to relax and make a little more flexibility about the way he handled the money for the troops. But Hinfpater stood on the book. And Franck said, "Herr Hinfpater, if you don't do what I say, I'll punish you. I'll drop a *pfennig* in your box every Saturday night." And Hinfpater blanched and went ahead with what Franck wanted. This gives a little idea of how Franck handles people.

Franck got cleared and became the boss of the chemistry division, succeeding Allison, approximately Christmastime, 1942. I don't think I intended to tell you what Franck told me about his relations with Seaborg, because Franck is still living. I'll tell you just part of the story. A very tense situation grew up between [them]—this is the time Seaborg tried to capture me and the part of my group that was capturable [*sic*] and for a period of time I kept detailed notes about the political action of the Project in my secret research notebook. I had, in the first months of the Project, done a fair amount of experimental work since I knew I wouldn't be able to do this long and I had to get some feel.

I first established—at the response to a suggestion by Lyle B. Borst—that the 12- or 13-day hard gamma species in fission products was barium-140 [decay]. I established—about the time of early 1943—that the gamma rays came with the [blank] [lanthanum, radioactive daughter of barium, from Lionel Goldring, May 15, 2010 email] and not the barium. As a matter of fact, I established that there were no barium gamma rays—that the barium gamma rays were less than 7% of the effective [gamma ray] counting [rate of], from W. Walters, May 5, 2010 email] gamma rays. It turns out that the right answer is almost exactly 7%. The sources available to me weren't good enough, and these sources I had created myself from cyclotron fission products—from the Chicago cyclotron or from St. Louis.

But back in the later part of this book whose number isn't known to me anymore, but it's in the Project records, is a day-by-day account of the fight with Seaborg without [detailed]—just high-spot statements. I kept these notes in January and February, and finally the fight got so complicated. We were holding our ground and there was just no point in keeping a record of what Perlman said to me on a certain day or what Hilberry said when I told Perlman what Hilberry said [that] Perlman said.

But I'll say just a few words because Franck told everybody that when Allison gave him the direction of chemistry he told him that his job was 90% psychiatry, 9% administration and 1% science. Then Franck would smile and say in his German accent, "But he overestimated the science."

Franck's scientific interest was very deep in the work that Milton Burton was doing in photochemistry and he met about half a day a week with Milton's men and knew them intimately. He made substantial contributions, through Milton Burton, and directly with various people. The rest of the time he was fighting off the Army, fighting off the physicists, and trying to prevent the chemists' fight among themselves from impeding the course of the work.

I saw him at least once a week. Sugarman and I would often go in together. He was not supposed to smoke, but he loved the little cigarillos that Sugarman used to smoke regularly. So each Friday I would bring him a cigarillo to bribe him to be on our side in the Seaborg fight for the coming week. This was the ritual that Sugarman continued when I went to Oak Ridge in September of 1943.

Indeed, when Prof. [Franck] was 70 years old—[he] may have been 75—there was a booklet in the late '40s given to him with statements from all his colleagues and I reminded him that we had bought his love with these cigars. But he told me, the first time I saw him as my formal boss, "I don't like this Seaborg, but after all I don't have to marry with him." A few weeks later I brought in to meet him a man named [Ermon Dwight] Eastman (1891-1945) from Berkeley, with whom Seaborg had done his Ph.D. Eastman would consult with the project at Berkeley though I don't think he was heavily involved. He died shortly after the end of the war. I just brought Eastman into the office and said, "This is Professor Eastman from Berkeley. Professor Seaborg used to work with him." Franck, without more than a how-do-you-do, said, "Seaborg puts every rock he can in my way and I put every rock I can in his." Eastman was horrified and Franck and Eastman and I talked a little bit, and I guess Franck and Eastman still more, but he thought Seaborg was the nicest person in the world and the most competent. After the Seaborg fight had died down, he said that there's no question Seaborg was the best [nuclear] chemist he'd ever seen and I think this is certainly true today. No one has come on the horizon since.

The last story I will tantalize the people who read this report with is when Martin Kamen came through Chicago accidentally—I knew he was strongly anti-Seaborg—he and Boyd and I got together one evening and brought him in to see Franck the next morning. Kamen hadn't called on Seaborg yet. Everybody had to call on Seaborg eventually but they often wanted to see the lay of the land before they did.

I brought Kamen in and Franck was tickled to see him and we talked a few words about Seaborg. Now Kamen is a very outspoken man, says caustic things with great wit, and few words were passed about Seaborg. Franck said, "Do you speak German?" and we all said yes, so Franck gave a characterization of Seaborg in German that is the only obscenity I ever heard from Franck. But it implied that he could still get along with the man. This was about the high point—this must have been the end of January of 1943—of the Seaborg fight. It was about [blank].

I don't know whether Franck will thank me for this but I can still say a few more things. The situation with respect to the DuPont Company eased greatly about the first of January, 1943, when Crawford Greenewalt came to the Project. The first DuPont representative, and a very likable, able person, was named Charles [Milton] Cooper, an MIT man. But Cooper had come to the Project probably just a few weeks after I did and had the problem of doing some pilot plant work on processes that had gone to Seaborg and Perlman—I think Perlman had been appointed liaison man with his section as Perlman had been appointed with mine. Perlman never did any liaison with my section. He helped me if I went to him, but he never made any constructive innovation of his own. But the other liaison man Seaborg appointed, Spofford English, was very helpful to our men particularly in instrumentation, where we were at our very weakest.

I think I mentioned in our first talk that I had never been able to hire a professional radiochemist, nor had we. We had two young men, Ben Schloss and Don Schover. Schloss had been a biochemist I'd known as a graduate student at UCLA. He'd gone into some Navy training program and learned electronics and this program had broken up and he had come to Chicago and was electronics man in our organization. He stayed in Chicago when the group split in September, 1943, and I took half the group to Oak Ridge. The other young man was a young fellow I think with radio training, named Don Schover, and he moved down to Oak Ridge with us.

Perlman had been involved with Bertrand Goldschmidt (I think I mentioned this before) in skimming the cream off the fission products without any information at all to us about it that led to one of Seaborg's setbacks while I was assigned to his group.

Cooper, presumably, had to have a process to go on. The only competence in plutonium chemistry was the competence of Seaborg's group in Berkeley and the greater competence he assembled in Chicago, and probably quite naturally, Cooper got a prescription (which I'd like to call it, not a chemistry) for taking neutron-irradiated uranium nitrate or neutron-irradiated uranium metal and somehow or other getting plutonium out of it without fission products and without other elements. The process that was being used by the engineers at this time was the lanthanum-fluoride process—LaF_3. This is the historic process by which neptunium and plutonium were both discovered, based on the fact that plutonium in its lowest water valence acts like a rare earth and will precipitate with lanthanum. Then the lanthanum fluoride can be dissolved in a boric acid and the plutonium can be oxidized to valence 6 where it acts like uranium. The lanthanum can be precipitated, leaving plutonium behind, later be reduced with some clean lanthanum carrying it out. This process worked rather well in a test tube and was the basis of most of the current analytical determinations of plutonium in the god-awful mixtures we have, for instance the stuff that my boys did in the first reactor, I think I also mentioned last time. But it apparently had great difficulties corrosion-wise and handling-wise in the big tanks and tubs that the engineers used in what was called the [blank] and to them was most unattractive to put in the anthra [*sic*] [lanthanum] plant we were talking about. Remember this $600,000,000 plant had to be designed for microchemical measurements on barely visible amounts of material and yet it had to operate on a ton-level to produce a kilogram of plutonium a day.

Franck was very uneasy about the fact that in advising the engineers, the judgment of Boyd, Burton and me and our senior men had in no sense been called on. The engineers were very uneasy that the chemistry department wouldn't take responsibility for the process that they were apparently committed to, and the source of this uneasiness at the ringside was probably Crawford Greenewalt.

Early in January (1943) he called me over to his office in [blank] and spent about three warm hours with me picking my brains for all I knew. Not politics on the Project, though I didn't hide it. I didn't press it with a stranger, but he knew there was a cleavage between Seaborg's group and the other three groups in chemistry. He never tried to exploit this and I don't recall ever talking with him in detail about it, but I'm not afraid to. It's just there were other things to talk about. I told Grace Mary that night that I met the brightest man I'd ever seen in industrial chemistry. He was bright enough to become president of DuPont without marrying the boss' daughter. I found out that he was still brighter, he was by blood a DuPont, and his wife was a DuPont. He later became president of DuPont Company on one of these two grounds.

He's now chairman of the board and I still think he's the brightest man I ever met in industrial science.

He created ideas—many, many Project ideas come from Crawford Greenewalt. And he had a warmth in handling human beings that was fantastic. It was also obvious the other DuPont people kowtowed to him a little bit later when I saw him in action. The party line from Wilmington made the same sort of kinks and contortions when we had his acceptance in Chicago and Oak Ridge that the so-called party line from Moscow does in Communist circles elsewhere.

Franck refused to give total chemistry support to the process. About November or December, 1942, Stanley [G.] Thompson discovered the bismuth-phosphate process, $BiPO_4$, and in January or February, 1943, one of the problems was to make a decision as to what would be the mainline process. As a matter of fact, I think that the bismuth-phosphate decision was made by the DuPont Company as late as May of 1943. But I remember, late in February, a meeting where the pros and cons of the bismuth-phosphate [and] lanthanum-fluoride process were discussed at a seminar where I had considerable acid comments to make because I knew from Franck how important this was. This was the last public discussion chemistry would have about the thing and, as I say, Franck had refused to sign the blueprints or to sign a statement that this was the proper method for DuPont to use. Franck had stood on the ground that we will state this is the best we know now, we must look at the best and the worst of these two processes and any others, including ion exchange that Boyd talked about. The DuPont Company was anxious not to consider ion exchange, probably for valid technical reasons. It was too new, but I think there were some politics in this also, and it was essentially a discussion of these two processes.

It's true that I was present at the big meeting in [blank], which I would state now to be early May, where umpteen DuPont men each got up and showed why the bismuth-phosphate process was perfect. The lanthanum-fluoride was not defended and there was a unanimous vote that the bismuth phosphate process be it and everybody agreed it would be easier. I think there were no strong defenders of the lanthanum-fluoride process but only the politics by which it was selected. As I say, this probably came about by the fact that it was selected too early and competence and judgment were too small among the chemists.

Franck said of Greenewalt, "I like that man Greenie, but I don't trust him." When he said he didn't trust him, it's quite clear to me and the other people with him when he said this that he didn't trust him in the sense that he represented an industrial force and we a scientific force and they would try to bum's rush us and bulldoze us automatically. The process was largely dominated by the deadlines of the DuPont Company. The DuPont Company would announce that we have to have the plans for a certain building, say the 205 building—the process built in Oak Ridge—they had to be final by a certain date and thus force all decisions to be made.

Now it almost always happened that when that date came the decisions weren't in good shape and it also happened that the building was generally delayed and other changes were made, but really the deadlines were encountered and this was probably appropriate. I've long since felt that it was inevitable that the chemists quarrel with the physicists. But this was totally friendly with total trust on both sides. I always thought it was inevitable that the chemists would quarrel with the chemical engineers and this was, in every area I've known, one without personal hostility— related to the thing. But the differences were deep and there was bitterness about the DuPont Company. And then it seems to be also inevitable that there be massive quarreling between the Army, insofar as it would interfere, and the Project. But for most things the Army was only involved in security and financing and purchasing and deadlines of big production equipment that I never bothered about, I never even knew about.

There was an Army officer, Colonel Arthur ("Pete") Peterson (1913-2008), who turned out to be a brother-in-law of Colonel (later General) Kenneth David Nichols (1907-2000), whom I didn't know was an Army officer. I thought he was just an engineer until I saw him in uniform at a New Year's Day party at the house of Kacy [Kenneth Stewart] Cole (1900-1984). He was with Greenewalt the day I had my first long talk with Greenewalt. But he was mainly an observer and never interfered with things I knew about of a technical nature. He did, however, with political problems.

Franck had the feeling that Arthur Holly Compton was too much of a boy scout about the Army. Before Franck came on the Project there was a lot of discussion about the—this was early summer of 1942—about our being put in uniform. I was a little opposed to this—I didn't know how to salute and I didn't know whether I'd get into trouble for sloppy dress in the presence of honest-to-god Army officers.

Compton seemed to favor this because it simplified the security problem, and we felt it would accomplish even more of a security problem. "We're secure, we know what we're doing, we're not going to tell people who are not supposed to know." Whatever ballots there were about this were not known to me in detail.

But in the planning of Oak Ridge we felt that Compton gave the DuPont organization way too much authority and Franck felt this very, very strongly. Indeed, in the planning of Oak Ridge, DuPont insisted this was a DuPont production site. A compromise was reached whereby the scientists sent by Chicago were paid by the Clinton Laboratories organization, not DuPont Company directly. We were not DuPont employees. DuPont probably didn't want the discipline of us and didn't want the commitment to us if the project failed or if the war ended suddenly and they had all these people on their hands.

The laboratory had elected Martin D. Whitaker (1902-1960) to be director of the laboratories, and Whitaker was a man I know very well and fought with a great deal. I respect the man personally, but I think he had way too little imagination and way too little courage. He was frightened to death of the Army and the DuPont Company.

His next-in-command, a man very close to him administratively, was Richard L. Doan, the man who hired Kimpton. I had wished Kimpton could have gone to Oak Ridge because I knew how to get along with Kimpton. I also knew how to get along with Doan, but I didn't know it for a long time. I may come back to that when I'm through with the Franck story.

Spedding felt very unhappy about Oak Ridge because in the early days of the crisis—in the peak days of the Seaborg fight—it looked to us who were anti-Seaborg and [*sic*] Spedding was certainly in this thing. Spedding had been displaced by Seaborg for six weeks. There was dirty work involved in this. Spedding had Chicago connections and was nominally in charge, under Franck, of Boyd, Burton, and me. He felt that Oak Ridge would be totally DuPont and Seaborgian, and he'd have no access to materials from Oak Ridge and ideas from Oak Ridge for his own group at [Ames, Iowa].

Finally, Franck brought this to an issue by getting some of the Project administrators—certainly Compton was involved in this, Doan was there, Franck, some people from DuPont and probably somebody from the Army—and there was an all-day meeting held in the suburbs of Chicago where Franck and Spedding batted it out with the Army about relations between Chicago and Oak Ridge and between the DuPont Company and the University of Chicago. And of course, Whitaker and [blank] were there.

Franck came back feeling that this had been a very useful fight and the right had been maintained that there should be no barrier whatsoever to communication between any Chicago group and its Oak Ridge sister. I had totally free access to correspondence and telephone calls and the like with Sugarman and, correspondingly, Perlman did with Seaborg. And the boys that Burton sent down which would be under my wing administratively, vaguely, were not responsible to me scientifically. They were responsible to

Milton Burton. And the people in Chicago had free visiting rights to Oak Ridge.

Whereas in other relationships, letters I wrote were written by me and signed Martin D. Whitaker, Director, and had to be forwarded to him—and this was very offensive to me. This was the DuPont-Army line organization. This is one of the things we fought. Now the keeping of free communications among these groups had a lot of good value from the standpoint of stopping the disease of compartmentation, which already had become overwhelming, and, secondly, of giving spark to small groups at Oak Ridge that might not have enough people or have enough courage to do their job right under the oppression and bulldozing of the Army and the DuPont Company.

When I moved to Oak Ridge I saw Franck a lot less, but I always went to see him when I went to Chicago and I went to Chicago—[because] there were Project council meetings and I would go to about two-thirds of these or send one of my trusted lieutenants to the others. And every time I was in Chicago I would see Franck. And any time one of my boys went up there, even if they didn't know Franck very well, I would just insist that they go in and make the acquaintance of him. And I guess Franck came to Oak Ridge only several times when I was there.

Franck visited Oak Ridge for the first time in early August, 1943. I may remind people that the date of the opening of Oak Ridge is the date of the opening of the cafeteria of the town site and to my recollection this date was about the 10th of July. Franck was kind enough to walk around the town and find out where the Coryell house was going to be so he could tell Grace Mary what size it was and was it pretty, because there always was a feeling that there were two kinds of houses in Oak Ridge—big shot houses that were lovely, and little shot houses that were in the boiling sun. It turned out not to be the case. Most houses were in the trees, and winding roads, and very artistic, but some houses unfortunately were in what they call the flats that were baking hot. The Coryells had been given a very nice house, a "B" house, the second smallest. But we were happy in that house and Grace Mary was grateful, and I was too, that Franck could bother to check up on what our house looked like and describe it to us. [As the house was a mile or so up a steep hill and roads unpaved, probably Franck surveyed Oak Ridge by car, not on foot. L. S. Goldring].

When Franck came down—I only can remember one visit because he also was not a well man. He had heart troubles and at the end of the war it was finally discovered he had ulcers also. But Franck still led me to borrow a car or bring ice—I had an old car by then—bring them up—he wanted to see Julie [Esther Coryell (1943-)], who was then just a few months old and say hello to Grace Mary before we took him over to Knoxville, to the train.

At the end of the war I had a chance to go to a number of places, including Chicago, UCLA, Caltech, Notre Dame, Washington University, MIT, as well as I had a strong offer to stay with Monsanto and go to their Dayton laboratories or stay with Monsanto and head up the chemistry division at Oak Ridge. I dropped all but the possibilities of Chicago, Caltech, UCLA, and MIT—MIT was the last big offer that I got. At the end of September, 1945, Grace Mary, Julie, and I went to Los Angeles and I did eliminate Caltech from the story. She felt that UCLA wasn't the place for me to go and I felt it was. But then I went to Chicago. I didn't want to go to Chicago on two grounds—I felt that if I went to Chicago the Seaborg fight would surely splurge up even though Seaborg was surely going to Berkeley. I'd be back in my old environment and I'd be like a warhorse on fire. And I had a great suspicion about MIT, that damn engineering school. I didn't want to go there with a bunch of conservative people. Whereas Chicago had a wonderful set of Project alumni coming back. They'd captured a good fraction of the brains in physics and chemistry from Los Alamos, Oak Ridge, Chicago, and Columbia.

So we went in to see Franck—in October, 1945, Grace Mary and I took our nineteen-month-old Julie in to see Franck in his office. I guess by then he'd withdrawn from the Project because he wanted to get back to photosynthesis. I don't recall who directed the Project. After all, the war was over and things were sort of dropping down fast in Chicago. The peak in Chicago had been past for some time. The peak strength in Chicago had happened in 1944. Franck had been succeeded by F[arrington] Daniels as director, and Daniels had the unpleasant job of seeing the Project decay before his eyes and trying to defend the integrity of people and the integrity of useful jobs against the exodus.

Franck talked with Grace Mary and me. He wanted me to come to Chicago. He told Grace Mary he didn't know why people liked me—it wasn't because I was so bright, but that Chicago seriously wanted me. Then he asked me to go out of the room and he talked to Grace Mary and told her not to try to interfere too much with where I went, that she should let me decide which was the most scientifically promising place. And he said, "Chicago is an awfully good place," and he put his finger up, "UCLA is an awfully good place, but don't forget MIT." This was the first powerful argument Grace Mary received for MIT.

I realized vaguely that Franck's two daughters had husbands in science in Boston. One son-in-law is Arthur [R.] von Hippel (1898-2003) at MIT, whom I got to know rather well since I came to MIT. The other son-in-law, Hermann Lisco (1911-2000), was associated with Harvard Medical School and is at present at the laboratories in Chicago in radiobiology. But Franck knew a great deal about the Harvard-MIT scene and the scientific eminence of the Boston area, whereas we West Coast people with the Caltech feeling against engineering way underestimated MIT and its participation in the cultural environment of Boston.

Since this time I've seen Franck very rarely. Whenever I've gone to Chicago I've seen him, except a trip I made very recently. He's now retired from his project. He married quietly in 1946, Hertha Sponer (1895-1968), an eminent German refugee spectroscopist at Duke University, and for a long time he worked at Chicago and went to Duke University at Christmastime and the two of them went for summers at Woods Hole. I've seen him at Woods Hole most of the summers, just briefly. I don't want press him. His health is very, very poor but his interest is high and his prime interest is photosynthesis and people.

I think I must say a few more things about Franck. Franck was very much concerned about the whole business—well, the topic will come up in his feelings about what should follow the war and the founding of the Federation of American Scientists and the like.

Franck was very opposed to the Morgenthau Plan to reduce Germany to a pastoral economy. He hated Hitlerism but he felt that you still had to recognize the dignity of the German people, and decency, and he went to Washington to express his views on this point. He also went to see Conant and Bush about postwar atomic bomb policy and atomic energy policy and he said that he never got such a quick, swift kick in his life as he did from Conant when he called on him, I think, at the American Chemistry Society building. Conant didn't want to talk about these things and he just flushed him out of the office.

He was asked in 1944, to go to Mexico City to celebrate the quadricentennial of Copernicus but the Army security wouldn't let him go into Mexico or leave the United States because they thought he might be kidnapped by our enemies and our enemies by that time, by implication, also included the Russians in the atomic energy field. Yes, we thought more of the Germans.

Well, Franck will come in from time to time. I may say that Milton Burton felt about Franck the same as I did, and when his wife had a baby in April, 1943, that boy was named Jamie for James Franck. He's now a student at Harvard [1963], freshman year. [Died 1987.]

Franck had the disarming way of handling chemists. Problems would come up (and Wigner had the same way and he used almost the same words) and Franck had simplified everything greatly and quickly and generally saw something fundamental all of us had overlooked in the mass of detail we brought in. So he'd be talking and he'd say, "Of course, I am no chemist, but what do you think of this?" and the thing would just come crystal clear.

Now I'm starting the problems of these rebellions. I'm proud to say I came in the Manhattan Project roughly six weeks before the Army was involved. My recollection is that General Groves came in about August 1942 [September 17, 1942]. We knew vaguely about the Army. I didn't know exactly how we were administratively related to other things until I saw the [Henry DeWolf] Smyth Report [*Atomic Energy for Military Purposes,* 1945]. It was just hard to tell. The relationships so much depended on personalities and accidents of meeting and the like and there was lots of intrigue.

General Groves was brought in to see the first speck of visible plutonium that had been isolated in Seaborg's group, and I think the date was August 18, 1942 [September 10, 1942], and I think the speck of plutonium is now in the museum at the University of Chicago.

Groves was contemptuous. This didn't interest him at all. We had to have a lot more than that. There was no thrill at all. Groves systematically took attitudes, made statements, that would antagonize people around him, make them work harder. I think he did this by design as well as by gift.

The chemist involved in this isolation was a very short man named Michael Cefola (1909-1983), microchemist from the group at New York University, who left the Project unhappy before the end of the war, and Lou[is B.] Werner (1921-2007), the tallest man on the Project, taller even than Seaborg. And Werner and Cefola were in a darkroom on the top floor of Jones in chemistry with their little microtechniques that the chemists had gotten down nearly pure. The Mutt and Jeff relationship was more pronounced even than Seaborg and Isadore Perlman.

The feelings about the DuPont Company that I discussed when we were talking about Franck not trusting Greenie were very right, and the dissatisfaction was at its greatest in physics. Sometime in the summer of 1942, and I would guess it's June, and I was at the house of Waldo Cohn, an apartment not far from the university and across the street from a park, where I met for the first time Arthur Roberts (1912-2004), the physicist who's done such fine balladry of science. He played some songs for us that evening. There was a man present named Lowan, Arnold Noah Lowan (1898-1962), I think. He was a physicist from NYU [New York University], but I just checked this morning in *American Men of Science* and he's had NYU connections but he's from Yeshiva University. He was on summer work in the Project—the first time I knew of someone ever being taken temporarily.

He fascinated me, and he represents as I recall it now, personality-wise the first example I've seen of a congenital radical, the New York ultraliberal Jew. He was griping with a rasping voice about the way scientists were treated at the bottom of the scale, we ought to have a strong union, it was a crime the way they paid us, and a crime in their decisions. All the things he said were right, as I see now, and I would have gone along with him. But I've seen it since the end of the war quite commonly in the left wings of science, but he was the first left-winger I'd ever seen.

I didn't run across Lowan very much more that summer, I don't know even with whom he was working. I don't know how Waldo knew him, because Waldo was in biochemistry and his other scientific connections were often closely with my own group. Waldo and I were like brothers from the time he showed up in the Project. He worked with Kacy Cole. But Lowan apparently had considerable influence on other radicals in physics, and since I'm in his group I can name them. I'm not ashamed of the integrity and loyalty of anybody in this group. But all spring there was ferment, and the ferment centered in a group consisting of George [Joyce C.] Stearns, who was personnel officer of the Project—one of Compton's former students who later died of cancer and who was head of the physics department at the University of Denver; Bernard T. Feld (1919-1993) had been brought into the Project at the time Szilard and Fermi were at Columbia. He told me just the other day that in Alice [Kimball] Smith's (1907-2001) article in *Harper's* ["The Elusive Dr. Szilard," July 1960], he was a refugee scientist [and] the only man who could bridge the gap between Fermi and Szilard, but unfortunately he was only a refugee from Brooklyn and he felt that she had not given him fair identification. The group also included Francis L[ee] Friedman (1919-1962), who is said to have been the first full-time employee of the Manhattan Project. He had been a young graduate student picked up by Gregory Breit, Breit being the Russian who set up these phony classification rules in good old Russian style (I think I kicked about them in my last discussion), [and] Norman Elliott, an older person, a chemist who'd been a colleague graduate student of mine at Caltech and whom I'd gotten to join me at Chicago from Pomona College. Now if I thought hard I could get many more names on the list. These names will come to me as time goes on. Yes, Alexander Langsdorf had come into Chicago. He was there but I can't date him on it. People I've named I know were on this.

Another leader was Philip Morrison (1915-2004) of Cornell, a man I loved very dearly and one of the brightest men on the Project, and a man that General Groves apparently trusted because already then I knew that he was involved in radiochemical counterintelligence about German aspects of the war. He made proposals about what we might find in German military equipment that would let people know German uranium and radium production, and the like, and he was an idea man all the way back. Of course, he was pilloried after the war by the Senate Internal Security Committee—had a hearing in Boston and he was professor at MIT and he apparently had belonged to the Communist Party in the late thirties and he and his wife had gotten out. But I have no doubt whatsoever about his total integrity and his total commitment to the war effort.

But the substance of this rebellion, largely in physics but with lots of sympathy in chemistry—we just had less imagination, less drive and we were more beholden to the DuPont Company and more slave to our work—was that the DuPont Company was holding the Project up by its luxury and by its red tape. And in time (this starts from discussions in the late spring and I'm now talking about what I guess is August 1943) decisions were made to: 1. inform Arthur Holly Compton of our rebellion; and 2. seek through Lowan a connection to Franklin D. Roosevelt and send him a letter asking him to send Harry Hopkins or Vice President

Wallace—we wanted someone whom Roosevelt trusted and who knew people and was one of the sort of emissaries Roosevelt would send to see Stalin or send to see England— to come and live with us a month and find out to what extent we were wasting time with luxury apartments in Oak Ridge instead of just shack dormitories, to what extent DuPont was holding us up—that they wouldn't let you process uranium until it stood 60 days, things of this type.

I was in a depressed mood all spring of '43. I found out later I had this psychological cycle, and the excesses of the boom in my fight with Seaborg in December, January, and February were followed by a depression that began in March or April. I felt miserable and I felt all my own shortcomings. I wasn't afraid of this fight but I didn't quite feel I had the knowledge or the courage to go to the discussions (there were meetings held). But finally Feld asked me if I would join them going to see Compton. I was busy, and I was interested. I had signed the petition (Norman Elliott signed in very big letters and said, "I'm the John Hancock. General Groves can read this without his glasses.") But on a day which is known from the records of Roosevelt's office as well as A. H. Compton's office, three or four people had an interview with A. H. Compton.

They were treated nicely, but Compton didn't let them talk very much. He spent most of the time talking to them, just filling in the time beautifully. They didn't know at the start of the conference—I don't know if they knew it before they left, from Compton—that that very morning Lowan had gone to see President Roosevelt. And, I think through [Robert Green] Sachs (1916-1999), probably, and [Albert] Einstein (1879-1955). He was in mathematical physics. On making the appointment a day or two earlier he told who he was and why he was coming. The President's secretary had checked with Chicago whether this was right and had informed Compton of the substance of his talk. Whether they talked or not I'm not certain but this is not a secret. I think he did have just a few minutes with Roosevelt.

Compton knew of the rebellion and knew he had gone to Roosevelt. He was a nice enough man to try to palliate the fight, but tried to deflect it without meeting its issues. He didn't give anybody hell and there were no repercussions known to me or anybody about this action, except the letter was given not to President Roosevelt but to General Groves and Groves gave it to the DuPont Company. Now if I know the DuPont Company it quickly destroyed the letter because the DuPont Company never liked to have fights in public. I saw Greenewalt lots the following fall. I never brought this up with him. I didn't know he was important enough with the DuPont Company. I just realized he was the brightest man they had, [and] they put [him] with us. It was after the war I realized he was the great power with the DuPont Company even at this time in these aspects of the DuPont Company.

It probably amused the liberal people in the DuPont Company and offended the other people, but the fight wasn't pursued further. It's my later and my postwar feeling that the DuPont Company did a superb job, but that we also did a superb job and these conflicts just had to be expected, and they didn't hurt anybody. But we felt pretty sick when we found that General Groves had our letter. But I don't know that anyone of the group that signed the letter—the petition was signed by probably thirty people. I signed early but the page was long and I think I saw it later when it was nearly full—I doubt that anyone was very frightened.

The second sort of unrest that happened on the Project I spoke about to some extent with Mrs. Smith, but I think I'd better rerecord it here.

Some time in March or April, there was a growing feeling of responsibility for us people to take the position of science in atomic energy policy seriously. So there was formed the Association of Chicago Scientists or Metallurgy Laboratory Scientists. This was a prototype of the Federation of American Scientists. It was formed somewhat in the spirit of the Association of Scientific Workers, an organization which had some red infiltration outside the Project, I've since found to my horror. The Boston branch of the Association of Scientific Workers seems to have been subject to Communist attack about 1939 and was captured by the Communists for a while and this led to some repercussions after the war. But that had no bearing on this, here.

Some people wanted us to be sort of a labor union, in the spirit of Lowan's discussion the first time I met him and almost the only time I met him, and some of them wanted it to be a high-sounding board for philosophy of atomic energy for peace, and some people felt we had to have some channel to form and transmit policy about how the bomb would be used. I had some misgivings about joining it—

JBS: Are you speaking about the spring of 1944?

Coryell: 1943. I left for Oak Ridge in September.

I had misgivings about joining it, which were based on two grounds. One is I was in sort of a depressed mood and this was messy and that I would watch and see, and the other is that I felt I understood so many of the young men involved in this—this centered in chemistry, not in physics—most of the young men involved in this I knew rather well and I certainly—I had no doubt in my mind about the radicalness of this but I felt that this would certainly make trouble with the DuPont Company and the Army and that if I were not a member, but friendly, I could be of more value to them. I also got this same sense out of Franck. They asked Franck to join and Franck said, "No. If I join you I will hurt you. I am a scientific aristocrat and I will do what I want to do and don't have to justify it. If I vote against you in any action, you're hurt by it, not me." And Franck told me not to join it. He thought I could do it more good by being a friend from the outside.

The president was, I think, Nathan Sugarman, who was my dearest and closest friend on the Project. Other officers included Truman P. Kohman in Seaborg's group—a man I got to know early pretty well and had a long and close association with ever since, a man not radical in any sense of the word but a man of very high principle and conscience, almost very devoutly Christian and with a sort of missionary feeling without missionary action. I never heard him try to persuade anybody against his will, but a man who knows his convictions of a high moral nature. Ben Schloss was another officer, and a young chemist who later became a biochemist down in Brookhaven, Dan [blank].

This thing perturbed the Army very, very seriously. Colonel Peterson was angry, and he told several members of this organization that if they didn't stop wasting government time talking politics in the hall, he would see they got to Guadalcanal. He'd draft them. It was a waste of government time, it was a violation of clean administration that these people worry about salaries, they worry about hiring problems, they worry about security, and it was a great threat to Army security if we talked about bomb problems. Our wives would get in on it and it would get all over town.

Sam Allison was furious about this statement of Peterson's. It may have been that this statement was made to Allison. I ran into Franck in Allison's office—I must have gone up about July, 1943, and this was only about a six-week-long period. There apparently were some meetings and some documents drawn up but I don't know about these and they may well have been long since destroyed. Allison told Peterson, "You'll draft any one of these men over my dead body. These guys have a right to think for themselves and act for themselves. I'll assure you they don't weaken security nor do they waste government time on this thing."

But Allison and Franck and Robert S. Mulliken—(Mulliken was an advisor to chemistry from the old days, called in really to help the psychotic Seaborg and Coryell, but who didn't have the emotional temperament to find out that there was a fight going on, probably, or what the fight was about. And I'm not sure how detailed Mulliken was involved in it. Mulliken was the information officer, you see, in a sense he was involved *ex officio*.)—I'm not sure Mulliken was present. I talked frankly for some time to Franck and Allison, and I think Mulliken, and possibly Hyman H. Goldsmith (1907-1949). He also was the information officer. He was a liberal not quite in the Lowan stamp, but almost that far out. He later was prominent in the FAS (Federation of American Scientists) stuff and at Brookhaven headed the information office after the war and was killed in a swimming accident in New Hampshire, leaving a wife and several kids.

I told some of the feeling these people had and the hostility to the Army, yet it was quite clear that Franck felt that the Project had to force the cessation of this activity. Apparently the conflict with the Army was too hot. And I would interpolate my own feeling that some of the men involved were considered to be unreliable security cases by the Army standard of guilt by association. I don't propose to name names. I don't know the membership in detail. But I think that the history of this short-lived association is one of the most interesting things in the Manhattan Project, and, as far as I know, until I mentioned it to Mrs. Smith it seems not to be known to the outsiders and is not known widely inside except at Chicago, largely in chemistry.

I was amazed to see Franck feeling that the thing had to be stopped, it would get out of hand, it would hurt everybody. So a meeting was called of its officers. Franck was frank with the officers. After all, Sugarman loved Franck as much as I do. I don't know why I didn't attend this meeting, but I felt it was not my business. I was trying to keep myself as a go-between. I think I brought information to Franck and Allison about feelings here that they hadn't otherwise had and I think some of these guys were going to break openly with the Army if they had to. I mean, there were some hotheaded guys who felt deeply about their principles. And this is not Red subversion in any sense—it's just people who feel that when you're going to kill 100,000 people you've got to bring some principles into discussion. The final execution may be by high-placed army officers or high-placed government officers, but they have to have heard the other side. Or when you're founding a new industry that presumably can save the world—and I'm speaking of atomic energy—you have to have some people of conscience and spirit who've thought about these things. And both these things are perfectly right.

Well, the meeting was held—and this is subject to verification, when and where—and several of the officers pointed out the problems that were being made of security and unrest. Finally, they talked with Franck and they talked with Allison. I think Franck was present and he spoke. And a motion was offered by somebody among the officers that the organization disband.

Anthony Turkevich, who was a member, felt bitterly that this wasn't right, and Turkevich was a man of conscience if I ever saw one, though not a bit radical in any other context. He's the son of an archbishop, now Archbishop Metropolitan of the Russian Orthodox Church, and he just believes in the dignity of man and he has this Russian argumentativeness. He'll just argue and argue to the bitter end. But so far as I know, Turkevich and two or three other people felt bitter about this and argued openly only when the organization disbanded.

It was sort of a cute joke, though. As part of the Seaborg fight my boys had to work very hard. I would go out and make messes politically and stick my neck out a mile and tell them all about it, and they'd have to do good enough work to show that we were as good as I was telling people we were. And we put out a major report (its number will come to me shortly or I can verify this in my papers at home)

115

involving all our work in the period of August or something like this, just before the split to go to Oak Ridge and some of us began moving in September, including me. The secretary, Susie Sedlecky (the first secretary I'd ever had, who was a very interesting, odd character and a very independent thinker also; she later married Turkevich) put down the title page which said Charles D. Coryell, Section Chief, and the next was Nathan Sugarman, Associate Section Chief, and I've forgotten who the other officers were – there were some group leaders, Rubinson and Elliott. She made a slip and put Nathan Sugarman down as Association Section Chief and Sugarman thought this was wonderful, a Freudian slip that proved that my secretary was on his side. She was not part of the organization.

How much longer shall we go on?

JBS: Well—

Coryell: Let's go through a few more rebellions.

I moved to Oak Ridge with Henri [A.] Levy (1913-2003), a relative newcomer to my group, a bright young man from Caltech. We drove down together and some other of my group came about the 23rd of September 1943. I think that's the date I took on new employment. I was a little bit annoyed because they cut me off in Chicago about September 20th, and I was paid travel money for the trip and probably $6 *per diem* but money was short in the Coryell family in those days.

Henri Levy and I drove down through Bloomington, Indiana and saw the cyclotron under a man named [Allan Charles

Gray Mitchell, CIT PhD, 1927, via emails from Guy Emery, Bowdoin College, and Loma Karklins, Caltech Archives, December 5 and 6, 2011], a man I'd known at Caltech and we spent a night at Denver [Indiana, between Chicago and Bloomington, or possibly one of two in Kentucky]. To our surprise, when we got to Knoxville, we found nothing but confusion. We had to go to the DuPont employment office—this was very offensive to me—and the office was a mess. I always thought the DuPont Company was efficient. I soon found out that nothing was efficient at Oak Ridge including the DuPont Company. They were the least inefficient of the lot.

We had to go get social security numbers. My social security number is one more than Henri Levy's. I was also proud it's all full of eights, it's fairly easy to remember, but I can't remember it right this minute. We were warned not to bring in liquor or cameras and I'm sorry that we didn't bring in liquor because they did not search very hard. I guess I got the worst search the day we showed up there. They did search a bit. We found out how bad it was to be in a dry county surrounded by dry counties. Thank God for the trips I made to Chicago where I brought liquor back for all my friends.

Vernon Cannon and Ed[ward Martin] Shapiro, Burton's radiation chemists who were now under me, had been down there some weeks already—as a matter of fact, had been there about a month. I had visited Oak Ridge for a two-day visit in August. I also came back and told my boys that this was wonderful in Oak Ridge, we'd have a northern enclave in the south. This was all going to be scientists from the north and we'd have a real liberal community. This was thrown in my teeth time after time until the end of the war because it turned out that there was a minority of northern scientists in a mass of southern workmen and there was nothing northern about anything other than the Clinton Laboratories chemistry and physics groups. The maintenance people and DuPont people were mostly from the south. The maintenance people, the project engineers, the employment officials—the security officer was a northern man, a real nice guy—most of the Army officers were southern and there was lots more Army control in Oak Ridge.

I don't think I've talked about how much influence Waldo Cohn and two or three of his young men had on me, and this is related now to the next rebellion. This is not a big rebellion, but it's a scientific and creative one. I'm pleased with my part in it.

We had won this fight ending with the big day Franck had, that we wouldn't be stooges for the DuPont Company, but it was obvious that we had to have many service facilities in Oak Ridge. Waldo Cohn is a young biochemist from Berkeley who had some time at MGH [Massachusetts General Hospital] and came to Chicago in December, 1942, and rather soon came to call on me. He'd only been around a few days. He was employed by Kacy Cole in the biology division to find out all about the fission products and their effect on animals. He was very much interested in techniques for radiochemistry. He'd had radioactive experience at Berkeley, I think, and at Boston. He had about 15 papers out. He was about the same age as I am, and had about the same amount of publications. He had been recently widowed. He had married a biochemist, non-

Jewish, who found she had cancer, and after she had known she had cancer, wanted a child to leave—a young boy named Mark. She had died when the boy was just a few months old.

Waldo came to Chicago with his baby boy of about two years old at this time—one of the sweetest little boys I've ever seen—and he immediately became close friends with Grace Mary and me. He was courting a girl by mail, who was also once divorced, whose maiden name was Eblin, Charmian Eblin. Charmian's father, whom I met later in the war, was editor of the Yiddish liberal magazine, *Vorwarts* [established 1897]. In later security charges involving Waldo Cohn, the so-called left-wing nature of his father-in-law's occupation was brought up, but Waldo can't read Yiddish and never read what his father-in-law wrote.

Waldo had met Charmian at a weekend party at the house of somebody in Boston and had seen her several times. They resolved to get married and they negotiated their marriage by mail and she joined him in January or February [1943]. (I don't remember the date now but this is also of record). Waldo immediately brought Charmian and Mark over to our apartment to meet Grace Mary. We were among his closest friends in Chicago, and they wanted me to be best man at the wedding. But I happened to be in St. Louis working the cyclotron, so they phoned me and Waldo and Charmian had to get married without my benefit. We've been awfully close to the Cohns ever since.

Waldo was concerned about getting quantities of radioactivity and he had with him several other biochemists. One of these was Paul C. Tompkins, who later was head of the U.S. Naval Radiological Defense Laboratory at Hunters Point, San Francisco, and is presently with the Public Health Service. Another biochemist from Berkeley, also with a tissue culture background but no relation to the first, was named Edward R. Tompkins. [Edward R. also directed the US Naval Radiological Defense Laboratory, according to ACS.org, and Paul C. became Executive Director, Federal Radiation Council, Washington, D. C., and worked on occupational health and standards.] They were called respectively, Red Tompkins and Black Tompkins, from the color of their hair, and they were very close friends to one another, not rivals in any sense. These three people kept pressing me to force the DuPont Company to give us at Oak Ridge a major facility to operate on a highly [radioactive] scale. Henri Levy, in my group, was also on this side. He'd come to Chicago in March or April of 1943 and was scheduled to go with me to Oak Ridge. The other people— some of the other people were interested. I'm sure that T. Harrison Davies, with a biochemical background, was interested. Harrison Davies, now one of the directors of the Mellon Institute, was an old friend of mine from Caltech days and deserves a little thumbnail sketch later, too. I'm sure he was one of the rebels on this Lowan thing… if he'd been there early enough. I'm not sure that he was.

In Oak Ridge, the DuPont people didn't want this. They felt that anything that took Coryell's attention away from the main line process of theirs, which was bismuth-phosphate, was a deflection from the war effort. But Waldo and I together generated enough enthusiasm in high circles—I think, largely R. L. Doan—that Whitaker agreed to send me to Wilmington. I guess the DuPont Company had veto power but wouldn't otherwise interfere, particularly because of this pact that had been signed in the Franck August meeting. They were very sensitive to being charged that they impeded scientific development.

Grace Mary had just bought a turkey for Thanksgiving, which she had cradled in her arms like a baby—it had been hard to get this thing and I announced that I was going to leave to spend my Thanksgiving in Wilmington, and she furnished the turkey to Vance K. Cooper and his wife, next door, and she had her dinner with them. Julie was very near to being born and Grace Mary was very unhappy about my going off, but it was for the good of God, country and the [father] land, and I went. I saw her brother and sister-in-law in Philadelphia. I spent nights in Philadelphia, I think, but I'm not sure, and I probably gave a lecture. I talked to most of the big shots in the DuPont Company, including Roger Williams (1890-1978), who was then president. They were very nice to me there. Greenewalt and I had long discussions about this and other things. Greenewalt was frank that they didn't want to interfere but, if they let us guys do all the research things we wanted, the deadlines would never get met and we'd never get done. And in a sense, I promised if they gave their assent to this hot laboratory, I would never pull my whole group into it and I would always put a certain fraction of my manpower aside for development on nuclear procedures for them, and in case of real crises we could do analyses—they had to in the division. This was all done in friendly teasing.

I also remember talking with Greenewalt at some length about the Jewish problem, and he said the DuPont Company wasn't anti-Semitic. "Why don't you tell me how many Jews you have in the company?" And he blushed and said,

"You're quite right. There are only one or two." He said, "We find they're trouble-makers. In other words, we don't hire people from California any more. You hire a Californian and you move him somewhere he's always unhappy and quits. So we hire, in a sense, to make a happy company."

When I came back to Oak Ridge, the first morning at the lab Norman Elliott, the hardest-bitten of my group, Harrison Davies, David Hume, and Henri Levy got me into the concrete-lined counting room, which was meant to be the counting room in our hot lab, and joked with me about bringing their bludgeons along to bash my head in. They wanted to find out to what extent I had sold them into slavery with the DuPont Company to get this hot lab. And we had a vigorous, hot discussion for about two hours until they were satisfied that what I had done was good policy.

In essence, my group and the piece that Sugarman had— well, it was about the same size—in Chicago, operated always on what I call Quaker democracy. No action is taken as long as there is specific opposition from any person, and it takes a long time to talk them down. I would, if I had to, pull rank in talking someone down, but I never did. And I was a little worried in going into this meeting, because I think this measures in essence some of the science-DuPont hostility.

In this same framework, a little bit later, there was a guy named Chambers, with the nickname Pop, who was a DuPont liaison man with the science group, a friendly guy who used to argue chemistry with me. And I was always contemptuous of the amount of brainpower that went into the procedures the DuPont Company used for their processes. They were totally empirical, as I see it from my side. During the winter of 1942-43, Davies and I first introduced the principles of freshman chemistry, the principles of mass action, into the DuPont process, the bismuth-phosphate process. It hadn't been done by Seaborg's men, either. And it's wonderful how far freshman chemistry will go. This gave me a lot of pleasure, to be a schoolteacher again.

But in the case of Pop Chambers, I agreed that I would do analyses up to a certain limit of so many man-hours a week if he would let me talk about the hydrogen ion. He'd resented this new concept from chemistry, a concept only twenty years old at this time, that the hydrogen ion in aqueous solution is not H^+, it's H_3O^+, that the pre-proton has so nearly zero a radius it latches on to a water molecule. Nevertheless, in thinking in these terms, certain analogies become apparent that don't if you have the old-fashioned, classical point of view. There was lots of teasing with DuPont people this way.

I don't remember having any hostility towards any DuPont man, specifically, but I had a massive hostility to the company. But I judged the men on their own personalities and they judged me, apparently, on my own, and I could tease them. We differed often and there'd be sharp words, but they would be the way scientists argue, which is rather rare in the DuPont Company.

I just thought of another one of these DuPont difficulties. The DuPont Company was already recruiting people for Hanford, and I don't remember the names of all the DuPont big shots that were in Chicago training, would visit Oak Ridge, and then try to get people to go to Hanford. I think I won't name the man in my group who was responsible for this thing. But one of my prominent young group leaders was badly needed by the DuPont Company, they said, and in fact they were right. They went to him to get him to go along with several others of my group who had agreed to go. One of my men who had agreed to go was Nathan E. Ballou, presently with the Radiological Naval Defense Lab but on leave for Belgium. He was a year and a half ahead of process chemistry of them all.

This particular man said he didn't want to go. A rather nice elderly man from the DuPont Company pressed him rather hard, and finally he said, "I'd rather be dead than work for DuPont." This burned the DuPont soul. I heard about this from Whitaker—that our men shouldn't talk that way to them, but I can't blame them. I felt the same way myself. I felt that DuPont bought the souls of men. They were very well treated salary-wise, they had nice working conditions, they had to hop to the line from Wilmington—to placate their wives, whenever they moved they got money to redecorate the house because theoretically the curtains and the fixtures and all wouldn't fit. The company paid the men it liked that had the right temperament, paid them plenty, but they were no longer able to think and act totally for themselves. They had to act for what the company thought was best for the company. This was the DuPont party line. I've talked about this with Greenewalt, also.

Well, the hot lab was authorized. I guess that already we had done some planning. DuPont Company furnished a man named Webb, Bill Webb, I think it is, who had been used to engineering things. He was a very competent man. He gave us all sorts of stuff we never thought about—how you plan washing out, how much space for moving carts around, how much wall space, and the like. But he seemed to have no love for the thing and he did a perfectly competent job. I don't think he's seen the building, for he's moved to Hanford. When I wrote it up for the National Nuclear Energy Series [Coryell and Sugarman 1951], I listed the detailed planning as having been done by Henri Levy, Bill Webb, I don't remember the third party—I'll substitute it in the record—and myself were the four chief planners of this thing.

Ground was broken in early December, I think December 6, 1943, right across from the chemistry building. The chemistry building was called 706A, the physics building was called 706B, and this was called 706C, the numbering system being a DuPont system whereby the numbers tell people that this is research. The number 700 is research, and 100 is production, and 200 is processing, and the like. Henri Levy and I were very close together. I think Waldo Cohn must have been that fourth man, obviously. The two Tompkins boys were involved in a secondary rank. Paul Tompkins had stayed in Chicago; Ed had come down. But I was amazed to see this thing built up. We'd worked out all the plans, they made sense to us, we had huge concrete cells in sets of four. I gave the name *bank* to a set of four cells. The cells were four-feet by six-feet floor space and eight-feet high inside, except one of the four was twelve.

[Audio file begins here, 27.06 minutes.] We weren't sure whether the operating areas in these cells would be along the sides or on the top and we had to make provisions for both. We also didn't know how we would see what was going on, but we made provisions for water tanks that we could fill with heavy zinc fluoride or lead bromide or lead salts, too—lead sulfamate [CDC spelled out]. Before we'd gone too far in the construction we found that an elderly physicist in Chicago, George Monk, would love to design us periscopes [*indispensable:* L.S. Goldring, September 28, 2011] and rather elaborate optical equipment was put in.

Another thing that Monk discovered was that the Army had designed a plane out of which the pilot couldn't see a full view, so at fantastic expense people had designed plastic lenses that would go between the struts and could see more than 180° vision. You could see a little bit of your own ears in these lenses—you could see more than half a sphere. Millions of dollars were spent on this and they were not used. The plane design was changed. Monk knew where they lay in the supply depot and provided us with a number of these. We used these a little bit. They gave what is called a pincushion effect. There was substantial distortion at the edges but they provided a method where we could look and see what happened everywhere in case of a crisis or an emergency. But I think they're not used in the hot labs. I'd love to have one of these as a souvenir even if it is scratched up and slightly radioactive by now. But we had to plan for flexibility.

But the shock was—Henri Levy and I, who almost always came in together in a bus, we walked in to our offices which were adjacent—to see this fantastic cement structure going up. It looked like it might be the high altar for a major cathedral—these huge blocks and blocks of cement, and then eventually a cruddy-looking building, a frame building, was put over it. This building cost about half a million dollars, but I think that this half a million dollars is not honest dollars. This was the last big DuPont construction project and with it, in the money granting, were taken care of a lot of funding for other things. Furthermore, R. L. Doan wanted to protect himself from the investigation we all figured would come at the end of the war for wasted money. Now it was often said that the minute DuPont came in we had no more to fear. If there were any heads to be scalped, no matter who made what mistake, the DuPont Company would get it. As a matter of fact, a man like Greenewalt would laughingly say this. So I don't know that Bill Webb had to sign as DuPont man, but Levy, Cohn, and I had to sign the blueprints to say that we felt that they were needed. This will be in the record somewhere—blueprints, just for historical purposes. I put a lot of love into this building.

The building grew rapidly. I think it was December 6th it was started, and by early February it was done. We had had in parallel—there was a big problem coming up. In the first place, this was built in case we would need it, but it was obvious to me we would need it and we had some ideas what to do with it. We had some kids working on techniques. Having the building, then you had to design apparatus that could make use of its strength. The idea was to be able to work in chemistry with very radioactive materials and without human exposure, and get things that otherwise couldn't be gotten in quantities available. But sometime late in February, Major General—I guess he was just a buck general—Leslie R. Groves came to the Clinton Laboratories unannounced, as usual, and told Doan and Whitaker they wanted to see the new building.

We were having a safety meeting that morning (this could have been March, not February). Safety meetings we had to have all the time. DuPont Company forced safety down our throat. We *hated* every safety lecture, we *hated* every safety poster, we *hated* every safety slogan. And this turns out to be the best way in the world to be safety conscious—anytime I saw something that would bring the wrath of the safety officer on my head, I took care of it. I still have a DuPont safety consciousness. This is a calculated hatred that's with me, a hatred of not being found totally safe. But we always tried to antagonize the safety officer by putting out safety-hyphen-science meetings, and we were talking about some chemistry which it would be necessary to be very safe in doing in this building and all of my men but one were there. The one guy couldn't stand safety; he was over blowing glass alone in the building. This was Louis G. Stang [Jr.], after at Brookhaven.

In marches Groves and Whitaker, and Doan, and a couple of colonels and here's a brand, spanking clean building and there's obviously not a thing going on in there but one lone man blowing glass. And Groves was furious. "It's a *waste* building this building when it wasn't in use." I got hell later on, but fortunately it was a safety meeting that kept us all away. There might have been four men in the building.

It also happened this building played a unique role in another side of the Project which satisfied my foresight and shows another piece of conflict we had with the Project officials. Sometime in November, Doan called up Perlman and asked him about the getting of a quantity of some gamma-emitter in the hundred-curie line. Now a curie is the same as a gram of radium. A hundred curies is a fantastic amount of radioactivity, never handled by anybody in the history of the world at one time. But our building was designed to handle ten curies. We had safety factors DuPont put in—it was designed for one with safety factors to ten, is the way I would honestly say. Perlman immediately passed the buck to me, rightly, and a meeting was called. This must have been about December 13th. I remember that date, but I'm not absolutely sure of it.

Well, Oppenheimer was inquiring about this thing. When I first heard about it by telephone late in November, I said, "Sure, it's easy. The answer is barium-140, lanthanum-140." The barium has no gamma (I myself had shown this) and the lanthanum had plenty of gamma—the worst gamma-emitter of moderate life and the lanthanum glows after you've got the barium done and the half-life is that of the barium mother, about 12.8 days. So you can do large-scale chemistry starting with thousands of curies of gross fission products in fuel from the Oak Ridge reactor, take the barium out without much gamma, and then between Oak Ridge and Los Alamos the lanthanum would go in, so that Los Alamos would get all the gamma they wanted.

Well, Oppenheimer came, I think it was December 13th. It was the first time I had seen him since Chicago days—it was probably the only time he visited the X-10 area, the plutonium project in Oak Ridge. We had a little discussion and he wanted to know if we would make about a quarter of a curie or one curie about March, and we'd make a hundred curies about May, and I said sure we would. This problem was obviously of high priority, and it tickled Doan that we could do it. It had apparently scared Whitaker to death because Coryell was so damned confident of what he and his men could do. But Whitaker recognized that it was an awfully good thing we had this building.

From that date on until over a year away, barium dominated my life. Because the commitments were easy, but these things had never been done. We just had provisions for them, the ideas for them, and this building was a unique facility without which the thing wouldn't work. I understand that when the idea was first suggested at Los Alamos that a radioactive technique involving the material be put to use, Groves was so happy he danced up and down. That's what Oppenheimer said in this meeting.

I drove back into town that night with the Army car with Oppenheimer, and Grace Mary was waiting for me to do her last shopping in Knoxville before Julie was born [in Oak Ridge]. I had no way of phoning her; we had no phone. So she was angry at my being late but she had her first meeting with Oppenheimer. She's never forgotten; he was very nice to her. He was tickled that Oak Ridge had the facilities for this thing. He thought he'd have to sell it or force it. He had the authority, could have forced anything down us as long as Groves was on his side, and from what he said, Groves was.

Well, we needed manpower. Doan was very friendly in all this thing, and pretty soon a group of eight GIs were assigned to me and they were all first-class men. They came on April 4th, 1944. The advance work was set up. Lionel S. Goldring (1922-), one of my former boys [from UCLA], and Louis [G.] Stang, [Jr.], and two or three other young guys set to dreaming up the techniques we needed. By March and April we were making runs, my chief lieutenants being Edward L. Brady, who is now at the U. S. Embassy in Vienna, chief scientist there associated with the American

[side] of the International Atomic Energy Association (IAEA), and Henri Levy. (I'll probably give more names later on; I have some of these marked down. I have some records now, who did what for a little while, but they're not with me.)

We sweated and sweated and *sweated* to get this quarter of a curie, first of all to make pilot runs and prove out all the components. We had the very great help of Miles C. Leverett [CDC spelled out], an oil company man [Humble Oil] who was the highest non-DuPont man in the engineering organization and he assigned a young engineer, Don Webster, who later went to Hanford and I think still is with DuPont. We didn't know beans about pumps and filters and all, and we had to start out on the hundred-gallon scale and end up on the milligram scale. To get a hundred curies of barium-140 we had to process ten or twenty thousand curies.

Well, this building had to have improvements made in it. We put up four inches of lead inside the cement walls to shield our equipment and we *had* to use our periscopes. We couldn't use any direct solutions at all. There was nothing but frustration and failure, and I guess I would have quit totally if it hadn't been for the stubbornness and sweetness of Henri Levy and the confidence of Ed Brady in spite of miserable failure. I also then knew how Charlie Cooper felt a year earlier, when every time they tried the lanthanum-fluoride process which looked simple in the books, how he felt, because it— like the barium process, looked simple in the books—failed time after time. *Fifty* times it failed. But Levy would always run a fifty-first and sometime or other, for a reason we never knew, it worked. And generally, when something once started working, then it always worked. It's the problem of handling trace chemicals when there was no engineering background for it, and little bits of impurities would stay in the stainless steel, or a little bit of welder's gunk, or a little bit of carelessness in making up reagents would lead to some tiny crystals that would take up this tiny amount of barium that we had. You see, we were dealing with a hundred pounds of uranium metal and we were looking for just a little tiny bit of a chemical that was moderately reactive.

Well, finally the pilot run that was promised in February or March was run in July. It turns out that Emilio Segrè was the man who wanted it. It turns out that he didn't want what Oppenheimer said, probably he changed his mind several times and somebody else changed his mind. I finally heard that when it got to Los Alamos it got broken, because it was in glass, as they had specified. It messed up the floor and they had to tear the floor out of the building and throw it away in a canyon. So all our efforts were for nothing except the training it gave us.

But it turned out Theodore [B.] Novey was going to seal off this barium. We worked about four or five days around the clock to get this one curie and it was *radioactive* as hell. It just beamed bad gammas. Well, we worked cleanly and quickly and nobody got any serious dosage. But then we checked back in the lab and found out that when you get lots of barium and you check, barium has plenty of gamma. But it's only 7% as bad as the lanthanum and I just found it was less than 7%, which is really thought of in terms of zero. I'm still proud that my early experiments weren't wrong, but they weren't very right.

Well then, I guess the hardest job we ever had was this quarter of a curie. Then we had to begin making timing with Los Alamos. We also got another big slug of GIs. We had to staff this building with a normal team of about eight people around the clock and still carry on the ordinary applications of my job. Everybody, in times of crisis, was yanked into the hot lab.

About the fifth, early in September, Richard Dodson, one of my dear friends from Caltech days, was sent by Los Alamos to be sure everything was all right. He and I were close friends and he looked this thing all over carefully. Later on, during the process, Gerhart Friedlander (1916-2009) was sent out. He was in Dodson's group. These guys had to handle the stuff when they came. I understand they handled it with not using massive cement, as we did, and remote control, and periscopes; they operated with telescopes at about 100 yards and had a fancy set of operations that involved puppet strings. They also relaxed the radiation requirements. We followed the rules, and they were not more than a tenth of an hour per day or six-tenths of an hour per week, and they allowed the guy to take two or three times as long, a half an hour a day or something like this. Of course, they knew the urgency better than we.

Also, Oppenheimer phoned me, after the run started, every day, and I had to go to the office of Whitaker and talk guardedly on the telephone and let him know how things were going. It was every day until the run started. I felt that if we didn't make our commitments on timing I'd be one of the bottlenecks in the atomic bomb. That's when I got to see how bad my transcriptions are. I was so nervous talking to Oppenheimer to transmit information without violating secrecy, and then I was allowed to see the top-secret transcription and I saw all my ohs and ahs and my bad English and the like, and I was very depressed. Also all the cusswords, all those things were there.

The run started. Whitaker came at midnight to see it—all the big shots were interested in this run. There was no question that it was hot stuff. We were thrilled the second day. The first day's output was about 40 curies. We had to build it up day by day, way past 100 to take care of losses, and then process it according to exact specifications and send a detailed top-secret telegram about every detail of the packaging. They furnished the fancy platinum rhodium packaging. There were supposed to be some GIs to convoy two or three cars, men with guns. We were sending a milligram of barium-140 with about a gram of ordinary barium surrounding it with about a ton and a half of lead surrounding it to make it tolerable.

It was scheduled to go Saturday, which would be about the 16th of September, 1944 [correct]. It looked like we had 300 curies. The sampling was awfully hard—to get a representative sample safely, and analyze it. You had to do it about a million times. Then we moved the stuff from the last engineering output storage into the detailed chemical process, and which with luck would only take about an hour to transfer and do, and none came.

It happened that it was Grace Mary's [30th] birthday [Thursday, September 14th], and I had come to work Saturday morning and resolved to just go back home and have dinner with her because this thing had to be seen through if it took the week. So I went home, I think on a bus, about 5 o'clock probably—a standard bus, and was going to have an armored car bring me back. We went to eat at the only decent restaurant—the north wing, which I call the north shore, of the main cafeteria—I think I met Jerry [R.] Coe's wife that night (he was later to be my boss) [James Robert Coe, called Jerry for pun on Jericho, Oak Ridge National Laboratory *Review*, Fall 1976, p. 44]. I went back home and there were three of my boys just frantic. There was no barium in the cell. Just nothing, well, ten curies.

I went back [to the lab] and we began chewing our fingernails and doing all sorts of tests to see. There were four cells, all contaminated, so you had trouble [telling] where the radioactivity was, there was so much being processed. The four inches of lead was on the outside, between the observer—there were observers, operating panels, cement wall, 2 feet 4 inches of lead, tanks, then two feet of concrete in the next cell, but the shine that came from the two feet of concrete would prevent your finding things in the next cell.

Don[ald S.] Schover came in around midnight and began putting together an ionization chamber we could drop down in. Step by step, every one of my boys were called in. We had some guys gotten at the movie, and Larry [Lawrence Elgin] Glendenin (1918-2008) was out with his date, and one by one, and an Army car went for the whole bunch. About 3:00 a.m. they were all there together and we were frantically doing everything when Glendenin saw the answer. In making the dilution a million-fold, the barium—practically none of it present, just a few billion atoms—all got lost in the glass. All the glasses were radioactive—they shouldn't be after you wash them out—so the simple answer was dilute with acid instead of water.

Then we had 300 curies. It had been there all the time, where it belonged, but we had no way of telling it. The riggers were being held. Now these guys, riggers, get well paid. They get time and a half for overtime and golden time for Sunday. They were called at 4 o'clock Saturday afternoon, though by the time they were called I knew we were in trouble, but I didn't know how bad trouble. Sunday afternoon they wouldn't come in or just would go away, but by Sunday afternoon I had been up all night. Whitaker phoned and Whitaker and Doan both came in. I should also have said that all this radioactivity in the cells and in the glass—you could just see it glow in its own light, it was a wonderful feeling. I remember in the movie of Marie Curie you could actually see the glow of radium, but we had the glow—radium glows better than lanthanum and barium, but we had 300 curies of barium and we had thousands of curies of other fission products. I was awfully groggy. I took a little cat nap, maybe, and Bob Lesch [?] was with me. We decided that what we needed was a swim. There was a water tank by the physics building. There was no "No Swimming" sign on it; nobody had thought about swimming in it. So we took a swim naked in the water tank, which did a lot of good for my morale, just as a contempt for the rules of the lab. The next day a "No Swimming" sign was put up, [laughter] and I got hell from the Army for swimming in it. I might drown, or the water might be biologically contaminated. I didn't give a damn. That did me a lot of good, just being naughty.

I guess Whitaker came in at midnight Sunday night, and about 8:15 Monday morning everything was all done. My secretary came in about midnight—this was June Babbitt, a wonderful girl. Doan got mad because you can't work women more than 40 hours a week and you can't work them nights, but she took a lot of dictation about all that we'd done. You see, we had long since quit taking records. This was all in my head. I was no longer stable enough to be allowed to turn buttons and valves and all, but I knew what was going on and I dictated all the things that the people know at Los Alamos—where it was distributed, how it glowed, what possibilities of impurities, what accessory tests we'd made. The boys went off. I suppose they drove day and night to Los Alamos. Then I know damn well that the Los Alamos people had just as much crisis and headaches as we had, because they had some chemicals just beyond us and they had less facilities and they also had practically no major tests on it.

When Dodson had been here, I took him to see Doan. Dodson came first from Los Alamos to do the last minute checking. Doan says, "I bet it won't work," and Dodson says, "Sure it will." Doan wanted me to just sort of tease a bit and probe how confident Dodson was of his work, because it was obvious there was a lot involved. So I took a bet of a nickel with Doan. I went to see him at 9 o'clock Monday morning, after I'd been up 48 hours and at work all but that little dinner, and there Doan had typed out a poem about me and my boys, and a nickel, which I'll bring to put in the record. I was never prouder in my life! Doan, a real tough old guy, he talked about "Coryell's chemists, why so flighty" and "Oppie blowing his toppie" to get this stuff and all. This nickel I gave to Lou Stang, who had done the key work on this. Stang's wife was very unhappy at Oak Ridge and insisted he move to Chicago, so as soon as the barium run was safely started before the night of the crisis, he had—he had also been working about 90 hours a week for about 5 weeks, he was no good at all to turn the valves and anything more. But as soon as he saw it start and thought it was all right, he went to Chicago. He deserved the nickel more than anyone else. So he has the original nickel and the original poem, but we had the poem mimeographed and given to all the 25 or 30 people who were involved in the process.

There was a meeting of the Project council the following Tuesday and Henri Levy, who had been up with me just as long, came to our house and had breakfast. Then he caught the train for Chicago to report to the Project council about the first successful production. Matter of fact, the last thing is—we thought we had 140 curies. When we got down to examining exactly how much a curie is, and Los Alamos reported back eventually through secret channels that we had 304. So we had three times the curies. There was the Los Alamos curie and the Oak Ridge curie, and nobody yet knows how many honest-to-God curies were in the thing.

This also came about the time of Rosh Hashanah and Yom Kippur, and the Jewish community had gotten some rabbi to come over for the banquet that followed the two days of fast. This came about September 19th or 18th, and Jack [Jacob Akiba] Marinsky (1918-2005) was the only devout Jew in the crowd. He went to the banquet and he was seated next to the rabbi, and he fell asleep at the table. He was sort of a shy guy, and he was embarrassed to death to think he fell asleep at the first Oak Ridge Yom Kippur, or post-Yom Kippur banquet.

It turned out Oppie had asked for one preparation of 140 curies, but already, before this was done, three more were ordered. This worried Whitaker no end. The building was being run on a crisis basis. I used to love crises; I didn't care.

Well then, also A. H. Compton asked me one day what was the upper limit we could do. Could we do a thousand curies?" I said, "Easy." Then he told Whitaker what I'd said and Whitaker called me in and gave me holy hell for over-committing the laboratory. "Suppose we stumble and don't do it. Then we disgrace the Clinton Laboratories and the Army branch in Oak Ridge totally." I said, "I don't care if we stumble and never do it. If we don't tell them we can do it, we'll never do it if they need it." I said, "I'll continue to tell people who come to me through channels if they are cleared, what we can or cannot do scientifically. It's up to you to worry about the administration." But this was part of Whitaker's fear—that things wouldn't look right.

Well, we finally got an order for a thousand curies. Meanwhile, though, there was big stuff now, and another building was designed under Miles Leverett. In the design of this building I was asked to be spokesman for the whole chemistry division about the chemical process. I gave a 10-page analysis. This is the most important technical decision I've ever made in my life. I gave all the arguments for and against the precipitation method that we'd stumbled on so many times, and I gave all the arguments for and against ion exchange. Then I came up unequivocally, 100%, for ion exchange. I knew perfectly well that the engineers would never accept it, that it was too radical for their ideas, but a couple of the boys of Leverett were siccing me on. Though ion exchange is now in use at the plant at Los Alamos that was built for this, and when I visited Los Alamos in 1951, I saw this lovely building designed for barium-140. They call it RaLa [Radioactive Lanthanum] at Los Alamos. This was then open to the public; it's not classified. Ion exchange is the main step. The first stage of separation, I think, is done at Hanford, and then they move it to Los Alamos directly. We took Oak Ridge slugs and later on Oak Ridge took Hanford slugs just to get more activity, because a thousand curies became commonplace and ten thousand curies is now considered the standard.

The engineers went ahead and they did all sorts of fancy things, and they drew all sorts of operating procedures and rules, and the engineers on the whole were contemptuous of the scientists. Now these aren't all DuPont people by now; there were some non-DuPont People. When they started their first run (we had done nine), they put a sign up—their building adjoined ours, abutted on it—"706 personnel keep out." They told us we would contaminate their building. Our building got contaminated now and again and we always had to clean it out. But nobody was hurt, and practically nobody even got overdoses in this wild operation. A few chemical engineers and GIs and a whole bunch of dumb chemists were running this show. When the fancy building that cost about ten times what our building did—this building still stands—ground through its first run, it ends up with a flat zero, because of a breakdown that was irrecoverable. And we had to go back in emergency operation to do the job. By then our dissolver was leaking. The apparatus had just sort of worn out, and this was a hit-or-miss affair, and we prepared over 1,000 curies. When that showed up at Los Alamos, Compton sent a telegram to Whitaker, "Orchids for Coryell and his boys," which we were told about and it made everybody feel good. The sign

"706 personnel keep out" came down when these guys goofed and we succeeded.

September 18th, for many years, I took a day off with whatever boys had been with my group at Oak Ridge—after the war there were Oak Ridge boys with me—as Barium Day. It was the biggest job I ever did in my life! [Laughter] Matter of fact, I think until 1950 or something, I was known to the people in the Project as the guy that prepared barium, therefore I must be a great engineer, and the guy that fought Seaborg. These are my two claims to fame.

The third, I'll talk about next time, is the element promethium, element 61, which Glendenin and Marinsky got done, barely, in the midst of all the crises of the hot lab.

Session Four — 12 August 1960

Coryell: Last Sunday you gave me the transcription from our first discussion, made at the end of that hectic luncheon after noon at the office [in Building 6-427] on July 14th. I had very little idea then of what I wanted to say and I have more now. The thing that worries me most is that I consider about the last third of my discussion rather unimportant, indeed some of it probably trash.

I find it very difficult now to describe accurately and to date with any accuracy my political feelings in the period 1928 to 1941—Pearl Harbor Day, let's say. I wasn't specializing in political science. I'd gone to Germany with practically no political orientation. I had the ability to observe but I didn't have the ability to think very effectively on what was going on, so I can see a confusedness about my thinking and there's an added confusion about my ability to identify my thinking at the time. Also, it seems to me that what I thought about Germany and things like—well, Germany perhaps less, but what I thought about war in general or European movements or liberal movements in the United States was based upon little experience and little judgment. I wasn't in any position of decision, and my opinions didn't affect more than two or three other members of my family or circle of friends.

On the other hand, when I came to the Manhattan Project I became a person of considerable weight because there were so few chemists on the project and I happened to be one of these few who had an independent academic background, and therefore had the responsibility to be a leader. I took this seriously, and I also by then learned that action is an important method of setting policy and I [blank] this. So many of the things I actually had bumped into or learned or made decisions on, I made even when I had imperfect evidence, swung other people and were rapidly built into my scientific and political personality, and made a difference in atomic energy developments, and made a huge difference in how I felt after the war, such that where I am now a fairly prominent spokesman for scientists in atomic energy, these decisions make a difference.

But what happened in 1934 or 1936 is still accidental.

The other thing is that I can date from my high school days when scientific or intellectual changes occurred in me, because I can date them with respect to the normal line of progress in the classroom—high school, college, graduate school, post-doctoral, Deep Springs, UCLA, and Manhattan Project—and essentially my major energies have gone into intellectual problems, particularly scientific problems. Now it sounds to me as if I'm making a speech now, so I've got to relax a little bit.

The ability to date things is important and I think my dating ability is fairly sound. School time is easy to date by the times of the year, semesters and the like, and when I was young it was interesting that my junior year and my senior year, and my first and third graduate years are big differences in my intellectual position, so I can date the year pretty well. And on the Project I can date things fairly well because so many letters had to be dated, and there exists, in classified files, a vast amount of correspondence if someone wanted to verify some of the things that happened. But I suppose better than throwing away this material would be just to leave it with mythology.

Where did we leave off last Sunday?

JBS: We left off with the discussion of the various revolts within the Manhattan Project.

Coryell: Oh, this was the second so-called "alarms and rebellions."

JBS: You ended with a long and very interesting story of the thousand curies of barium.

Coryell: Did I tell about my visit to Wilmington to get the DuPont Company to authorize—I think I did.

JBS: To authorize?

Coryell: The Oak Ridge branch of the DuPont Company indicated a definite lack of interest in approving a hot lab and, I later heard, because they felt it would be a diversion of the work, not only of me and my group but of other research groups away from what the DuPont Company felt was the main line job—namely, to polish the bismuth-phosphate process to perfection—into fundamental chemistry that obviously wouldn't shorten the war. I don't know how much I—

JBS: You spoke very little about that. You spoke about the—

Coryell: I discussed going to Wilmington, I think, because I now remember telling the story when I teased Greenewalt because he had no Jews in the DuPont Company. I want to be sure to get that on the record. I think Greenewalt would enjoy it because he was utterly frank with me about the matter. I might say that I have been in the chemistry department at MIT for 15 years, and not a single Jew has been appointed in my tenure there. It's not my fault. It's a slight discrimination. There are always several men interested in a position when one comes up, and the powers-that-be in our department apparently unconsciously have selected a non-Jew over a Jew of the same ability level, or will accept a sunny, open-minded person, rather than a morose, self-centered or surly person. And so we get sunnier people and less imagination, less conflict.

JBS: Do you suggest that MIT is run on the same basis?

Coryell: No, I just think the chemistry department is run on the same principle. The physics department is probably half Jewish. The area of [nuclear] physics is heavily Jewish. It's interesting, in respect to the Jewish problem—which will come into each of the discussions one way or another since I think it's a very important problem in modern times—some areas of science appeal to Jews or get a few Jewish leaders. They then will appeal to many more of the younger students, as it's easier for them to get in. I still think a Jewish student has a harder job planning a career in American society than a non-Jew, and I think a Protestant has a better chance of finding a career than a Catholic. There are natural discriminations, I think, within the human race. Each person likes someone of his own social class, orientation and temperament—which almost always has about it an element of the nature of the educational system the man [!] came from. Public school people dislike private school, and I'm in this category. Athletic, vigorous Rotary Club joiners like people of that category and discriminate against the eggheads.

In the Jewish situation the discrimination is said to be based largely on manners, and there is a definite class of manners that comes from the Jewish immigrant who had to work hard as a boy in the slums of New York to come to the top and tends to be extra-combative and aggressive, or the children of these, which have a somewhat different personality type but generally have been pushed by their parents so they have this same aggressiveness, but they have different goals. So one can mark Jewishness by manners, and by accent in the old generation, often by Jewish names (though more and more names get changed), and discrimination goes on.

Well, I'm probably talking the same thing over several times. I want to talk about several fields of science that are heavily Jewish, and one of these is theoretical physics. Another one of these is nuclear physics. I can't explain, I can't put my finger on the exact way this entered, except that it is true that because of the tradition of the Jewish race, the Jews have a heightened interest in theoretical problems. Because of the nature of Talmudic training [Wiesel 2009], the Jews that have had this have a heightened interest in hairsplitting, and these both have some effect on the type of study a field will take up. There's a legend the Jews are no good as farmers, and no good as soldiers, and no good as experimentalists, but it is a standard Israeli joke, "Look at Israel. They're no good as businessmen in Israel, and they're good as soldiers and farmers."

At any rate, in the American educational system, which is not terribly hairsplitting and not terribly intellectual compared, say, to the British, the Jewish mind, which begins in devout families with heavy Bible training, and a good deal of training in Jewish lore at Hebrew school on Saturdays, just tends naturally to go to the philosophical problems and the metaphysical problems and which theoretical physics makes a modern scientific bridge to. Well, then, when three or four prominent Jewish people enter a field it's natural that Jewish scholars will gravitate to them. When job placements come along, it's natural that the leaders in the field will recommend able people they know of, otherwise unemployed, and so the Jewish contribution to the field is thereby enlarged.

Now the depression years were awful years to get jobs, and Jews had a much harder time than non-Jews. Except that a discriminating school that wanted the very best young man in theoretical physics or theoretical chemistry could, by taking Jewish candidates, get a substantially better man than the [blank] at the time.

Well, now to get back to the Project (I'm still talking to the machine rather than to you), the Manhattan Project was the last of the major war projects to form. I was brought to Chicago in May of 1942, and the only good scientists I knew of not pretty well occupied in war work, were Jewish. This is the residuum, the leftovers. And this is also true of the physics projects, though nuclear physics hadn't been involved in any important war project. But in general, physics had been. There were less people available for nuclear physics, but on the other hand, it was very easy for the leaders of the Project—Oppenheimer, Compton, Lawrence—in locating men, to persuade a man with a nuclear background that he should leave his job at this time and come to the Project. It was very easy for a nuclear physicist to get wind, a nuclear physicist could easily get wind that there was a major nuclear project. It involved fission, it involved chain reactions, and they would gravitate by word of mouth to get in because the security wall wasn't tight and it's a good thing it wasn't. We'd never have been able to hire.

Well, when I got to Chicago in May of 1942, I had tapped to bring with me several young boys. One of these was a man named [Edward L.] Brady, whom I hadn't known to be Jewish. That is, I hadn't known until I had known him for a couple of years. His father had been an Abromowitz, but his father had decided to change his name so there wouldn't be discrimination against his children getting jobs. Brady never hid the fact that he was Jewish, but never volunteered the information unless he was in a Jewish environment. I never inquired, and still never do, whether a man is Jewish or not, and I'm sometimes surprised to find that some people are so bright I'm sure they're Jews—they turn out not to be; some people are awfully dumb and turn out to be Jewish. Jews have a better record than Gentiles in grades, because they are under more pressure from their families to make up for the social deficit that their Jewishness brings with it. But it's a very tragic thing to see a Jewish family pressing a dumb boy or a boy who isn't happy in the profession they've chosen for him. He's under tremendous pressure from all the uncles and aunts and brothers and sisters to succeed, and it sometimes causes breakups.

Well, when I began to look around outside of the boys I brought with me, almost everybody I saw who wasn't occupied in an important way who looked like he'd be easily makeable into Project material was Jewish, and I had this affection for Jews and interest in the situation, and I began to learn a lot about West Side Chicago Jewish life and South Side Chicago Jewish life, and some New York transportees had come in. There were pressures. I found that older Jewish people have a pronounced anti-Gentile feeling when it comes to the intermarriage of their children. I was horrified at the fuss raised by a nice old Jewish couple when their son decided to marry a non-Jewish girl, even though this girl was prepared to embrace the Jewish faith. And I understand this more knowing more about the Jewish—

JBS: (Inaudible).

Coryell: Yes. It's Jack Siegel's pop and mom. But I think Ed Brady's father and mother would have had the same fuss.

JBS: You said that DuPont had, if not a spoken policy, at least a general policy against hiring Jews. Was there any conflict at all in Oak Ridge about the number of Jewish scientists from the Manhattan Project who joined the—

Coryell: There was precious little anti-Semitism within the Project. We were so anxious to get the work done, and the field was dominated by nuclear physics which, I would say, was already probably 40% Jewish, and many of the leaders were so obviously Jewish you didn't argue about it. And most non-Jews, like me, when they got close to the issue realized that these were awfully good people.

I didn't finish my story about—I mentioned how the Jewish families object when the children marry Gentiles. I also found that a Jewish family is very warm to any Gentile they meet—that is, members of the family meet—if this Gentile is obviously open-minded about Jewishness, and even still warmer if this person is curious about Jewishness. They'll talk about it, they're glad to talk about it. They don't understand the issue, either, and as a non-Jew I could often point out things that the Jews didn't see about themselves, both good and bad, and help explain to them the funny characteristics of white Protestant society which is dominant in the United States. They're eager to talk about it and I've always been eager to probe it out, so as I'm doing now, I brought it up.

I mentioned, I think in my first recording, my rationalization about why so many people charged with subversion are Jewish. It comes from their early preoccupation with the Hitler problem, where they were shoulder to shoulder with Communists, and partly from the fact that the Jewish uneasiness and the Jewish theoretical hairsplitting lends itself to Communism. As a matter of fact, in high school or early college I worried about the problem that the anti-Semites say that the world bankers are all Jews and the world revolutionaries are all Jews. These are all expressions of Jewish training and Jewish willingness to work hard, one oriented to making himself rich, like Bernie [Bernard]

151

Baruch (1870-1965), and the other oriented to making himself wild, like [Leon] Trotsky (1879-1940), who is the symbol of Jewish revolution I had in mind there. But there's an eminent Jewishness about both these people, as far as I know from all I've heard about them.

Oppenheimer said he didn't discover his Jewishness and what it meant until the Jews were under heavy attack from Hitler, and this is why he gravitated to left-wing groups who were already largely Jewish and were worried about this. Many Jewish people—there was a policy in Europe and a policy in the United States of assimilation, and this is what I would have thought—and almost would think today—would be best for Jews. But when I got to Israel I realized that the stubbornness of the orthodox Jews, and the stiff food customs, and the stiff marriage customs which, we would say, press to mark the Jew from the non-Jew, have kept the integrity of Jewish culture at the expense of millions of Jewish lives. The Jewish community of the world, now numbering 15 (?) million, paid six million lives stubbornly fighting Hitler as Jews, when many of these people could have escaped the thing by migrating to other countries and merging with the dominant group. But I think that the Jewish culture, which underlies the Christian heritage and the Western heritage, is an important one and I'm glad it survived the Roman Empire and the fall of the Roman Empire by this same stubbornness. I'm glad it survived the fall of Hitler, and it could survive a maelstrom of East-West war, I think, by this very stubbornness that will cost lots of lives. I think it is a tradition worth maintaining, but I personally would hate to pay the price by being an odd ball with a long beard and hair at the corners of my head, the *payot*, that the orthodox Jew wears, wear a hat all the time in the house and even to bed, which the orthodox Jew does, and refuse to do a single lick of work on Saturday, including lighting a fire or turning a light switch on, but this is the routine that develops from the Roman time and the middle ages to retain Jewishness. But it's not so much that it was commanded by God in the Bible, as that it's been proved by 3,000 years of history as a way of maintaining a way of life.

Well, I think we should get back closer to the Project. I was going to give some names of people who joined my group, but maybe the best thing is make this as an appendix. I can write to you at the end of this whole program, after I've read the dozens of pages, because it turns out that lots of people who were young have since become very prominent. The Project had such an emotional impact, such a scientific impact on the literally thousands of people who worked on its research phases that this will bear fruit in the scientific world for many generations. After all, an academic generation is only, say, ten or fifteen years—the time between a man taking his graduate degree and a man producing graduate students in quantity—and it will affect more than the lifetime of all the people on the Project. I was just fortunate to come on the Project at an age of early maturity, so I got into more things, and I also came as a chemist when there were so precious few chemists.

The physicists felt they knew all the chemistry they needed, and the engineers thought they were chemists. So that I found myself in a very prominent position and this led, at the end of the war, to seven offers of a full professorship in important places in the country, and associate professorships in still more important places, whereas the average physicists of my age and my productivity, having a field that was already very large, were (all) getting assistant professorships at the same level. I think [Victor Frederick] Weisskopf (1908-2002) came to MIT as an associate professor, and Francis [Lee] Friedman came as a graduate student—he hadn't finished his doctoral work. Bernard Feld came as assistant professor, and I came as a full professor [1946]. I am embarrassed. I'm not on a par with any of these people, scientifically, but there were at the end of the war only six or eight chemists who knew much about the Manhattan Project who had announced that they wanted to be nuclear chemists. Some of the older men, such as Farrington Daniels, whom I've mentioned as the last director of the Chicago project, had a lifetime of photochemistry and related subjects. He continued to add atomic chemistry to this and be interested in the peacetime applications, but research in his laboratory was not seriously modified and he didn't take students who would want to be graduate students. Warren Johnson was the Director of Oak Ridge who got appointed [September 1943-December 1945]. He was an inorganic chemist of distinction but already was pretty much a fulltime administrator. He went back to Chicago at the end of the war as Chairman of the Chemistry Department and I think he's now dean of physical science. He never had a graduate student since the end of the war.

There were a few chemists who called themselves nuclear chemists before the war and they went back with distinction as full professors. These are Willard F. Libby, who had been at Berkeley as associate professor, probably. He went to Chicago as a full professor. Glenn T. Seaborg, who had been at Berkeley as an assistant professor went back as a full professor. Joseph W. Kennedy, who had been an instructor at Berkeley—the one I tried to hire for UCLA I mentioned about—who went to Washington University of St. Louis as a full professor and department chairman; Arthur C. Wahl, who had been a graduate student just fresh post-doctoral at Berkeley went to Washington University, I think as an associate professor, at the end of the war. I think that that covers the list, though I shouldn't say this because I don't know the whole story about the Columbia SAM project.

Then there were a group of others, like me, who had not been in atomic energy before the war, but had some academic position, who went as full professors to universities if they went. I'm sure that George E. Boyd at the Oak Ridge laboratory got offers in this regard, but the University of Chicago I think only offered him associate professorship. He stayed in Oak Ridge. Clifford [S.] Garner [Chemistry and Metallurgy Division, Los Alamos] went as an assistant professor to UCLA, though I had been offered a full professorship there. This gets a little hard without a list of names.

JBS: These were people who were young enough to change fields, essentially, to become nuclear chemists during the—

Coryell: Nobody's too old to change a field. I change it every ten years. I think that some of the older men like Daniels and Warren Johnson—and there must have been more in terms of the eastern project and the Columbia project—didn't work in such central positions in the plutonium project. I mean, they were doing the chemistry of uranium. They went on and did some more uranium, but they weren't nuclear chemists in the sense [blank] nuclear physics, as I count myself, partly. Or they weren't traveling salesmen for nuclear chemistry and for radiochemistry who drew applications of radioactivity to ordinary chemistry as I've always been. [CDC prided himself that he was professor of chemistry in broad sense.] I've always been a soapboxer. I learned this during the war, that advertising is an important part of science. Besides me, there were very few others with independent position in a university who were called and, therefore, the number going on to full professorships like me was also very small. But there were people such as Sugarman, who I consider better than I am and only a few years younger, who went to the University of Chicago as assistant professor. Well, he'd never had an academic status, so he didn't get a higher rank. The guys who got full professorships in chemistry would have been professors when the war started. As far as I now can recall, I'm the only one in this category, but there were assistant professors like Seaborg who went to full professorships, but they were in the field. This made it even better.

But there were lots of younger men who had never had academic position and had PhDs and they went in as assistant professors, but rather rapidly moved up in their departments, and I think they gradually stayed in the field of nuclear chemistry. Then there was a very large body of young men—and this is the sort of thing that ought to go in the appendix—who hadn't finished their graduate work and elected, therefore, to go to one of the centers like Chicago, St. Louis, Washington University of St. Louis, MIT, Columbia, or Berkeley, and get their degrees. These are now the great body of leaders in nuclear chemistry in the country today— they're numerically a large number and, I'm proud to say, that of course, many of these are boys who worked with me during the war. The University of Chicago took a large number of these and let them include in their thesis work relatively declassifiable parts of their wartime work. Matter of fact, these theses were classified in most cases, but they could accept problems that were clean enough and enough done by the man alone to be counted. It wouldn't be a part of a big group. There were enough people at the University of Chicago who had the clearance of the Atomic Energy Commission that the theses could be discussed widely and defended in seminars, and the like.

MIT took the boys that came at the end of the war with me or Professor John W[ithers] Irvine, Jr. (1913-1997), and let them use part of their Manhattan Project work that had been done under my observation or Irvine's right at the end of the war, and they'd add to this a little bit and get a thesis somewhat easier. They were proven high ability and ability to do research, but they wanted enough of this done at MIT so that professors who weren't in the Project could judge for it and MIT, rightly enough, wouldn't accept anything classified in a thesis.

It's possibly worthwhile for a digression on the problem of classification. I think that the classifications during the war were fantastically narrow-minded, and a damaging sort of story, and I think I'll come back in a minute and tell how I got involved in this. But I think that classification in a university is damning. It damns the university department that classifies theses to mediocre work, because another result of having classified work in a university is that the whole staff of the department can't participate and it can't be freely discussed with members of other departments. It tends to set standards where the university is nervous if there's a man who is pronouncedly anti-government or suspected of not being clearable. They therefore tend to squeeze him out and discriminate against him in discussions in seminars and the like. It means that the man who is doing a thesis in a classified area only can discuss it with a few people and tends to feel the rule is you don't discuss it with anybody unless you have to. The net result is that work is not fully judged, the work is not thought about adequately by the person doing it and the person directing it, the work isn't defended against competition adequately, and the new result will be both mediocrity and a divisiveness in the department with respect to the gravy train boys that have classified projects and the others.

Now, MIT must have classified theses, I suspect, in naval architecture. But the department is so nearly totally cleared, and this is such an ancient problem I suspect it's handled quite cleanly, but I think no other department—well, perhaps electrical engineering today, some problems in guidance and the like—but they only do this because a huge body of people may do it very carefully and they do it in such a way as not to harm that small fraction of the department that isn't in the clearance family, or the foreign post-doctors, or the foreign staff members.

Chicago did this thing I talked about with theses in a generous way and they didn't invite new classified work, they invited the use of this work with the understanding that it would be declassified, but I suspect that these theses have not specifically been declassified. Maybe they have.

JBS: Do you know of any specific instances at Chicago where the classification led to the easing out of somebody? You mentioned this as a problem were it any university.

Coryell: It happened in one department at MIT. I can't remember the department, and I can't remember the name of the case, but it concerned me very much about 1947. It must have been naval architecture. It was a department where every single member of the department had traditionally been cleared, including foreign members as a matter of fact. They had a Dutch man, Trost, as chairman up until his retirement. He was appointed about 1950. But he was an eminent man, he'd been in the Dutch underground, and presumably he had an informal sort of clearance suitable to the United States Navy. But there was one man who had clearance problems. It was normal to ask for clearance for every American citizen and, I suppose, to negotiate some suitable sort of analog of clearance for the non-Americans. It was decided by some government agency that he was probably not clearable and they ought not try it because if you get a flat "no" it makes more trouble than if you don't. He was under some pressure from the department to withdraw in a way that wouldn't embarrass him and wouldn't embarrass the department. He chose to argue about this. He came to see me and I worried a lot about this, talked to three or four other people. I don't remember the name and I wouldn't put the name on the tape [recording] if I did, because I think this is the sort of error which should be kept confidential until the man himself wants to speak because it's a serious detriment to a man to have him marked as unclearable unless he cares to make the announcement and give the explanation that makes sense to him about it.

As to Chicago, I was talking about the general principle of dividing the department. But I know how our department felt in the late forties when the United States Congress decided that people getting Atomic Energy Commission fellowships would have to take a loyalty oath. This was contested by the university as much as present universities are contesting a loyalty oath for students who get federal loans. Harvard just refused to take this money [as did] a number of universities. Apparently MIT is equivocating on the subject. It hasn't been talked [about], to my knowledge, at MIT publicly. I'm concerned about it, hope that MIT takes the Harvard position, but it hasn't been a live enough problem that I'd contest it. But in the late forties this was contested by the Federation of American Scientists and the Associated Scientific Workers, by prominent people at Harvard and MIT. It's the only time I saw joint action by prominent chemists at Harvard and MIT in a political matter, because the chairman of the chemistry department at MIT is, in essence, apolitical, wants to keep out of messes, and he doesn't interfere with other people's activities that they see fit, but he likes to play a very cautious game. But this apparently became of enough concern to the university that a number of chemists from Harvard and MIT met together at MIT. [Blank] was then at Harvard—he's now with the Bell Telephone Company—was the leader of the discussion. Jerrold [Reinach] Zacharias was quite active, [as were some] prominent professors at Harvard. There was essentially a representative number. I won't name names because I can't guarantee that everyone I think was there [was there]. I haven't any reason to think it wasn't essentially a group opinion of the Harvard department, because there was such a good cross section of it, and the same was true of MIT, including [Arthur Clay] Cope (1909-1966), whom I had previously thought was backward in seeing the political implications in academic life. Cope was one of these who explained to me first how MIT felt about classified theses. This is apolitical.

JBS: I think, if we could, we might move ourselves back to 1943, which is where we had logged and move ahead with our—

Coryell: Alarms and rebellions? Well, I'd like to continue in the classification context. I told you on my first tape that—I think that was the tape where I discussed how when hiring was done, we started in the early days to be very frank with a man we felt we could trust and we were very liberal in saying who we would trust. I would trust a person trusted by any person I trusted, and I think that's the way it ought to be. There was this conflict between telling a man all he needs to know to do the job, and no more than he has to know. There's no question in my mind—you tell a man all he can understand and you make a better man out of him, he'll do a better job, and I felt in 1942 the war was a long and dirty war. We had to spend the time and money on education, because I knew I had to learn everything about nuclear chemistry starting from nothing. I learned this roughly in six months, and I wasn't any better than the young boys I was hiring though I had a better background of chemistry to build on because I was older.

I also essentially made this assumption: these guys can do anything I think I could have done at my best at their age. I feel that this policy brought the best out in my men all the time. Men that I now think, and thought then, would have been mediocre or middling good were superb. Ed Brady's in this category. He had trouble getting through a master's degree because he had a personality conflict and a philosophy conflict with his professor. I had been sympathetic with him. But he had a rather good record, was willing to work hard, but in this war atmosphere he just flourished. He was challenged every minute, every hour. He was struggling to do better than himself. The competition in universities that makes good students is often that one student is jealous of another and tries to outsmart him.

Well, here we were—there was little interpersonal jealousy, but we were all trying to envelop and overcome a tremendously difficult problem calling for all possible skills, namely, the founding of the nuclear age, the building of atomic power, and the building of atomic bombs, and just about any skill that one could think of had some bearing on this. Anyone who was confronted with these problems, of the brilliance of Fermi, or the breadth of Franck, or the breadth of Arthur Holly Compton still had thousands of facts to learn and hundreds of principles to worry about and put together. It was just so utterly new and it went all the way from principles of astrophysics, fusion, reaction, fission, chain reaction having analogs in astrophysics—down to biology for the protection problems and the national standards and the wartime standards and the radiation that might be. Biological warfare was talked about, and the like; to some extent agriculture, but certainly chemistry and applied chemistry of all sorts.

JBS: Did you at any time run into any opposition to this policy of informing people as much as you thought you had to?

Coryell: Anytime I put it in a letter I got hell, because it would be [go] no place. The letters to people outside the

Project had to go through the director's office. I tried to hire one of my best friends at UCLA, a very bright and warm man, somewhat older, who had been somewhat bypassed by the war projects. This is E. Lee Kinsey (1903-1961). I had reason to think he'd come, I'd spoken to him over the telephone with some broad hints and he said he was interested, write him a letter. So I wrote a letter about it, and I didn't want to say anything that was violating security, but I wanted to pin it down and I wasn't very smart. I started to tell him how happy I was emitting neutrons and gamma rays. I figured it would hint, that if I talked this way that I was concerned about the theoretical problems on this line. This letter got scotched, and I got holy hell from the security officer and from the director of the laboratory. Being somewhat chastened, I don't think I wrote him some sort of perfectly dull letter, and there was nothing very commanding about it in spite of the fact that I made a phone call with some broad hints. Kinsey never came for an interview directly, and we lost what would have been a very good man. As far as I was concerned, had he been hired by our group (which had physics concern) and put in our group, as he would have liked to have come because of his personal ties with me, we would have had a pet physicist who could educate and enliven us much faster than we learned this from our dilettante contacts with the physicists for which we were sometimes pet chemists. I felt very badly about this deal, but it was partly my own ineptness at handling it.

We had a series of classified training lectures. I've mentioned the Seaborg lectures, I think, which were held at Chicago. There was a series that was held at Oak Ridge, which have just come out in a government report. I might say, in a footnote, that my chapter was written in great haste. It had to be written up after the lecture and I made a mistake in the relative [value] for electron masses and it stands frozen for all time in a training book. Just some lousy physics I put out. There was no machinery for correcting it after I realized it was going all over the DuPont labs.

Also, later in the war, there was always a hunger for education. After all, the Manhattan Project in Chicago and the Clinton Laboratories, the X-10 Project at Oak Ridge, was heavily university-oriented and university-led. The DuPont men who were best at leadership also had this same hunger, and the DuPont people, in general, would gladly go to see what these wild scientists really thought. You had to come up, slowly and coherently with an outline and some mimeographed notes afterwards, but here it was bubbling about [blank]. Later in the war, and I would hesitate to date this sharply without a lot of thought. I'd date it with the birth of the Nordheim baby [Erik Vincent], probably, so it must have been spring of 1944. There was a series of lectures more of the seminar type, given in the Oak Ridge school— this was the Cedar Hills school near our house—but everyone had to show the badge which marked that you were cleared for the research area. Various badges showed different things. A plumber who was cleared for that area could have come, and he certainly would have been welcome, but as I didn't hear about it I doubt that any came.

In this series of lectures, I was asked to give a lecture on the fission products, which was my major obligation, and I tried to make this lecture as stimulating as possible so I talked about the lovely things one can find in fission products. Among these are the element 43, technetium, and the element 61—the element 43, which was once named masurium but it wasn't really identified, and the element 61, which had once been named illinium that wasn't properly identified. Element 43 is now called technetium, in a post-war decision about names. This comes from Emilio Segrè, virtually, and element 61 is called promethium by suggestion of my wife, Grace Mary. After we spent about two hours one night trying to explain to her how important an element was, it took her ten seconds to realize the beautiful statement in [Percy Bysshe] Shelley, *Prometheus Unbound* (1820). This was so pertinent we were delighted and we named the element after the Titan.

[In 1995 co-discoverer Larry Glendenin quoted the conversation: As Grace Mary was a poet they went to her with their story and request for a name. Her response was, "I got a 'D' in high school chemistry. Refresh my memory, what's an element?" To their explanation that an element is irreducible matter, she said, "Name it after Prometheus. You have stolen fire from the gods, and mankind may suffer for it."]

I think my lecture went over rather well. I think it was rather interesting to see that chemistry does play a part in this role. A few nights later I stepped on the 5:30 bus, the late bus in the laboratory. As I stepped on there was a person in the driver's seat, two men in separate positions far apart in the bus whom I don't remember recognizing, and Mrs. Nordheim, Gertrude [Poeschl] Nordheim (?-1949), the wife of Lothar [W.] Nordheim (1899-1985). I sat down in front of her and started talking. I knew her very well, she lived near us and I liked her very much, and she started to talk with pleasure about the lecture I had given. I felt uneasy because the lecture was classified in the sense the subject matter was classified and fission was in this area you couldn't talk about. But I couldn't stop her. And then she said, "What was most interesting was what you said about illinium." And I thought, "Ah, ah, she's getting specific." I had two choices which came to mind. One, I should tell her that this was a classified subject we couldn't talk about, which would immediately draw attention to what we'd said. The other was to laugh it off and tell her this was old stuff and name some pre-war date, which I did. Then I turned to the *Knoxville Journal* paper, and began reading funnies like mad.

I forgot this, but about two weeks later I received a phone call and was asked would I come up to the administration building about a quarter of a mile from the research lab with Warren Johnson, the director of chemistry. I said, "Sure." We were planning a new hot lab. I went on to Johnson's office a few minutes after 11. He'd heard about the thing but didn't know what the topic was and presumed it was hot lab. We walked over cheerfully.

I walked into the office of R. L. Doan, the associate director of the laboratory and the director of research, and he had the longest face I've ever seen in my life. As soon as Johnson and I had arrived, Martin D. Whitaker, the director of the laboratory, walked in, and his face was even longer, and in walked a man named Murphy—Edward [J.] Murphy (1900-1965), I think, who was the highest Army officer at the Oak Ridge National Laboratory, and he was sort of worried about security and the like, but I knew him well as a chemist and was a good friend of his. And he had a very long face, and he rather seriously said he had a very unpleasant task, he had to inform me that I had been heard talking on the bus on such a date about uranium, and that if this were true he would have to evict me from the Project for a major security violation.

I remembered clearly everything I'd said to Mrs. Nordheim, and I just said, "I wasn't talking about uranium," and I got truculent. I told him exactly what I talked about, and that I had to make a security decision, and that there wasn't anything I could see better to do than what I did. It's to deflect this into prewar dating, into an interesting piece of chemistry but not very important, and drop it. So Martin D. Whitaker felt he had to give me a fifteen-minute lecture about what a major security break it was if it were known that a prominent chemist with an interest in the rare earths (which I never had had) was in Oak Ridge working on element 61. It obviously means we're working on atomic energy, it also means we're close to having a bomb, the timing of the bomb is the most important secret of the war. This was all just utter nonsense and I said so, courteously but firmly. The whole thing was a misfire. I hadn't said this and I felt somebody was snooping. It probably was a good thing they snooped, but I knew my ground and I walked out.

I felt I would never speak to Mrs. Nordheim about this because she, too, would have gotten hell and, if we ever came up again, and knowing these goddam security people it was likely to come up again, I wanted to be able to testify honestly that there had been no collusion among us. So only after the war did I share stories with her. But I found out from Doan, the day he left the laboratories. I came over to bid him farewell and expressed my regret he was leaving atomic energy. I also bet him two dollars he'd be back in the field before two years were up, but it took him about four, so next time I see him I've got to pay him two dollars. We've seen each other once since he came back and I forgot the bet. Doan told me he was hoping that he would fire me because he thought it would be the best thing in the world to have a major fight about firing Charles Coryell. He would find I had many friends, he knew I had many friends, and it would help clean up all this nonsense. But he was also proud of the fact I knew my ground and stood my ground. He also told me that Mrs. Nordheim had been questioned ahead of me and that she had taken all the blame. She was in a moderately advanced stage of pregnancy and they wanted to go very easy on her. I now realize that she remembered the story though probably not as sharply as I had because she didn't feel as intense about it as I had. It can also be said probably, there were people who would like to get rid of me, and Doan knew about this, and they figured this security thing would be a good basis to get me over a barrel and there would be other things they could call up in the deal. I would love very much to have been able to have seen the security dossier on me, because as you can gather from the last discussion, I was in almost any fight that came into sight.

This is what I learned from Franck without Franck ever saying it. Franck always spoke out for the right whether he could win the fight or not, and I decided that under war pressures and the like there were dangers of accidental misfirings and failures and bums rushing and it was necessary to be frank. Whenever I thought something stank, I should say it boldly and move in. This is quite a difference from the wishy-washy period in the thirties when I was equivocating about what was right and wrong. On the other hand, I'm not sure I would have been—with my 1945 wisdom—a better man in 1935. It's one thing for me to stick my neck out and fight with Army security when the major risk to myself is embarrassment to myself and family, eviction from the Project and drafting into the Army and going to Guadalcanal. It's still a personal sacrifice I make, a sacrifice of myself and my immediate family. But if I were in a position of authority, say, in the Atomic Energy Commission today or in the Department of Defense and I had to make a yes or no answer which meant the difference between wiping out half the United States or not, I still think I'd be wishy-washy. The order of the magnitude of the burden of deciding for right and wrong when the cost may be 10,000,000 lives or 100,000,000 lives, or even 100,000 lives is quite different from deciding just for my own career.

Now, Grace Mary was very uneasy about these things. We never talked about atomic energy—she never had been frank with me about what little she knew about it, which would have relieved me a bit because it would have taken the edge off my conscientiousness about security—but she knew I was in trouble at the lab, off and on, and she'd hear from other people about their delight about how I stood up to the bosses, or their worries that I wouldn't, or that somebody else had and sure as heck I'd be in the fight next. She always exaggerated the dangers of these things and she also tended, as a woman, to think in terms of herself and her baby and the terrible disgrace if we were kicked out of the Project. We would be slackers or worse. She thought in these terms rather than if one were kicked off the Project one could be proud of it.

Well, I think that you like rebellions so I'll try to dredge up a couple of others that occur to me as I go along.

Matters of Justice and Policy. And I'll preface this with a statement that I heard Eugene Wigner make on December 2, 1947, at the Oak Ridge High School, at an open meeting celebrating the fifth anniversary of the opening of the atomic age. The atomic age is considered to have opened the day of the first reactor in Chicago, [Stagg Field, University of Chicago, December 2, 1942], and I don't think I told you the story of that day, either. I'll come back to that another time. It isn't a rebellion.

Wigner said that one of the great difficulties of the wartime Project and the postwar project was that major decisions have to be made by people who belong to a very limited group of cleared people, and it's necessary for a democracy to have enough people involved in decision-making to be representative of the people for whom the decision is binding. It's furthermore necessary, for the furtherance of the human mind, that important decisions be made by enough people that you can really kick the ball around and have the yeses and the noes and the alternatives fought out so you can make a wise decision. One man can make a fast decision, one brilliant man can make lots of fast decisions. But one brilliant man can't stay perfectly right forever because he can't contain the quantity of information and the quantity of alternatives that a group of ten or twenty or thirty men can. This was one of the serious problems of the whole Manhattan Project.

On the other hand, many serious policy decisions were taken in a vacuum and were never applied because so often in the

Manhattan Project scientific policy was set not at the meeting of the Project council in Chicago every month—as a matter fact they never had time to make decisions, they never had time to finish an agenda of being informed—decisions were taken by the action of a man who was excited about a new finding, or two new findings—one that came by word of mouth from Los Alamos and one that came from the laboratory at Oak Ridge—which would get together at Chicago. A bright idea would carry the Project forward whether it was decided to go ahead or not.

In the closing-down days at Chicago there was a manpower limit that was decreasing month by month in the various projects and there was a conflict between Nathan Sugarman, my Chicago twin, and Thorfin [R.] Hogness (1894-1976), who was under pressure to cut down the luxury of research that the Sugarman group was involved in since the wartime problems of fission products were mostly transferred to me and to the group at Hanford under John E. Willard, and all mostly done. A decision was made which Hogness had to transmit to Sugarman, as Hogness was director of chemistry and succeeded Franck sometime after I left Chicago, that they should cut out the research they were doing on gaseous fission products. The time I'm talking about is about September 1943, or October. (I can date this if I have to with some books I have at the lab, [specifically] the Coryell and Sugarman dates, the Plutonium Project Reports on which research these 1951 [336 wartime] papers were written. These carry dates with them.) [Coryell, C. D. and Sugarman, N., editors, *Radiochemical Studies: The Fission Products*, New York, McGraw Hill, 1951.] Sugarman was very unhappy about this and resisted vigorously and cheated anyway. He couldn't resist if in the normal course of an authorized experiment with a little bit of addition you can find some unauthorized data, he couldn't resist finding it. I'm 100% on his side.

They were studying something relatively well known, 9.2-hour xenon-135. There were some good glassblowers there. Louis Stang, my bright boy from the hot labs had gone to Chicago and was in this group, but he was marked to be dropped because he was one of the latest comers and the group had been cut to eight men and he was the ninth in Sugarman and Rubinson's idea of who was of the most value for a dying project, to do some research and be able in a crisis to activate their old lines of work.

Suddenly—I guess I was in Chicago on one of the Project council trips which [blank], a phone call comes from Hanford through this classified teletype with a suitable amount of garbleness, and indicates that the big Hanford reactors are suffering from sleeping sickness. Obviously I'm not talking about 1943; I'm talking about 1944. You turn the reactor on and run it a while, and turn it off, and you can't put it on again. It happened that John W. Wheeler was at Hanford as physicist in residence. The DuPont Company, besides not having Jews, only had one and a half physicists, so it isn't so bad. They had a number of physicists they added during the war and John Wheeler was one of these who was very helpful to them. What had happened was that the nuclear pile, the reactor, created by its own fission products, poisons. It suffered from auto-intoxication. Given enough time, the poisons would be eliminated. The elimination process is radioactive decay. Wheeler was able to make a fairly good guess from the timing relation of how long you ran before you stopped, how soon you could start again, and the—I think he measured the excess reactivity in the reactor, and this looked like some 6-hour species, which was not a poison, was creating a 9-hour species which was a poison. Fortunately, Sugarman's men had just completed a detailed study of the 9.2-hour xenon-135, which was the most likely explanation for this.

I called Oak Ridge and in garbled terms, so no one would guess what I was talking about—we had code names for the elements and code names for the numbering system, prearranged. It was quite a transparent code in that the initials of the elements were named by cities. I think Constantinople was cesium, and I can't remember what we did for xenon right now, but we had some way of handling it. Maybe it was Xenia, Ohio. There's a list of this somewhere in my notes, classified at Oak Ridge. Norman Elliott immediately realized that if this was the case, if xenon would stop a reactor by absorbing neutrons, you'd put some homemade xenon into the reactor where there were lots of neutrons, the xenon will get wiped out by the neutrons and the effective half-life will be shorter in the reactor than out, and within hours Norman did this experiment to verify it.

The other big question they needed to know at Hanford was, what is the fission yield of the species? How much of it is produced, and therefore what change do you have to have in loading pattern for the reactors, and what changes does this make in the production problem? Sugarman, in violation of the orders that came down from the Project council through Hogness, had enough information—this could be pieced together with relatively high precision and with assurance— for Chicago to tell Hanford that if it were iodine [decays to] xenon, this is the story. And my boys at Chicago could tell Oak Ridge, "honest-to-god, this is xenon."

Now at Oak Ridge we had a very liberal policy. Richard L. Doan promised that 10% of my resources could go into scientific research on anything I wanted. By fighting hard I kept 10%, but the hard time to fight for this 10% independent research was at the very same time, in the fall of 1944 when we were working like devils on the barium job. Norman Elliott was so contemptuous of large-scale experimentation, and so contemptuous of the DuPont Company, he successfully evaded any obligations. He just made himself so nasty when brought into the hot lab, we never brought him in, and he was able to be free for this problem in a very creative way. Glendenin and Marinsky were so hot after the element 61 that they were dreadfully unhappy when they were impressed into the hot lab. I mentioned how Jack fell asleep on the rabbi's shoulder, but as soon as he wakened he got back into the research lab. Jack would do anything I asked, and Glendenin would do anything I asked with a little pleading. So I, who would have liked to have been 10% of my time in research, too, was zero percent, and was of practically no value to these boys while they were finding promethium. But they had the drive that they could work half-time on this problem—because there were lots of guys not working any time—and we still had less than 10% research. But some of this research was rather fundamental in science, which added to the position of atomic energy after the war, but almost any of this stuff found out proved to have other values. It happens that promethium-147, the 2.7-year promethium they were working on, not only is the first representative of element 61, but it represents a matter of some significant toxicity among the fission products and this needed to be known in detail by Waldo Cohn's boys and the like. And furthermore, it is now used as a major power source for the SNAP Project [Systems Nuclear Auxiliary Power], and things like that, and these things were already being talked about in the old days, and this information was worth getting, even in wartime heat.

Now, in a sense, one of the deep differences between Seaborg and his operation of his group, and my operation of mine was he had an oligarchy of the bright boys and they called all the shots. Now they had very democratic discussion, but the democracy was four or five men—the group leaders who met every Thursday night—and the boys down the line were told what to do and nothing more. There was a much slower time that the brand new boy with a fresh bachelor's degree in electrical engineering or biochemistry or chemistry or something like this came to be in a productive position. But in my smaller group, where everybody was just—we put pressure on them to learn and all decisions were made in my group by Quaker democracy. No action was taken unless all opposition was beaten down in open argument and I considered it immoral for me to call rank on these boys. Once in a while there'd be an Army decree I'd have to enforce, or once in a while there'd be a problem that did have a Los Alamos classification I couldn't discuss widely. Then I'd bring in my group leaders. But there was no problem I knew about, that I had to be concerned scientifically about, that I wouldn't share with my group leaders although technically I might have been violating some security. I just wouldn't trust myself to make the decisions alone. I found that the collective wisdom, added to mine, of Harrison Davies, Henri Levy, David Hume, Ed

Brady—the irradiation boys, Edward Shapiro and Vern Cannon, were interested in only part of this stuff but when it involved irradiation chemistry or something they would have [blank], I count them in as my group leaders. I was responsible for my group, but they weren't responsible for everything else going on in my section. But this collective brain power was used, and, of course, this is the group of boys who took me into the hot lab to beat my brains down if I sold too much of our soul, collectively, to DuPont in order to get the hot lab. Waldo Cohn wasn't a member of my group but he was almost always in stuff like this that involved operations that he had experience in. I considered my group pretty general and I sort of played godfather to loose characters running around that would need an administrative home or could bring us stimulation or some idea of what problems might come from....

I was awfully close to Sugarman in the time I was in Chicago and leaned on him tremendously, emotionally. We collectively leaned on Rubinson and Elliott—Brady was one of the kids, but one of the oldest kids, the first to grow to group leader. He didn't have a doctor's degree, but he had been in the Project almost from the time I came, and he had a maturity and judgment that made him, late in the war, rank equal with people with PhD and three or four years post-PhD experience, but less Project experience than he.

At Oak Ridge, Waldo Cohn was certainly the closest to me of anybody not in my group.

JBS: Did this independent research which you did—or which your people did—get them into any kind of trouble at all? Was there any complaint?

Coryell: The DuPont Company was contemptuous of the fact they never could rely on us to do work because we were always interested in our own work. DuPont Company's sole interest in life was taking the arbitrarily chosen main line.

I'd like next to tell the story of my knowledge of Martin Kamen's security problems. I met Martin Kamen first, before the war, as one of the bright boys at Berkeley and a close friend of Ruben's. Martin Kamen and Sam Ruben discovered carbon-14 at a very interesting time in the project, and just in time for E. O. Lawrence to use the information in his Nobel Prize [Physics, 1939] award address [Wheeler Hall, UC Berkeley, February 29, 1940]—although he had opposed the time being spent on it for half of its development because of the pressures on him in the laboratory.

Martin is a completely unorthodox, extremely brilliant, sassy, snotty, short man with as much love for baseball as chemistry, and as much knowledge of physics as chemistry. He's presently in biochemistry at Brandeis University. He was totally anti-Seaborgian from the start because he's anti-stuffed shirt. I liked him independent of any knowledge of the Seaborg situation before the war, when I was at Chicago, only twice that I recall now—once, when he was on his way east to see Urey. He arrived at the same time Kennedy did and there was a private meeting at our house which was a stop-Seaborg meeting. The next morning we were going to George Boyd's house, since George Boyd was a great ally of mine in the Seaborg fight. There was a deep snow on the ground, and as we were going down 56th Street we saw some of Seaborg's boys on their way to a meeting. We had to hide behind cars so it wouldn't be detected that Martin Kamen and Joe Kennedy were in town. They would make their perfunctory calls on Seaborg the next day, but they didn't want him to know that they were [blank] through me.

I also found out that he had a tremendous capacity for liquor, because we tried to get him drunk that morning to get him to talk more. He was talking [taking] all we could handle. At a later time in Oak Ridge when he visited in February 1944, I took my precious supply of [blank] whiskey and tried to titrate him to the end point. The whiskey gave out when the fifth was gone, there was no more whiskey and Martin Kamen was still in good shape.

Martin knew a good deal about the plutonium project because he was bright and he had seen this stuff developed and he kept his ears open. He was a chemist at Berkeley, working on problems of Y-12, which is mostly dull uranium stuff he hated. He was utterly contemptuous of Army people, particularly security officers, and utterly contemptuous of officials of the Tennessee Eastman Company. He arranged, in one of his visits to Oak Ridge in February or March 1944, that I come over to his laboratory to talk to some of his people about uranium-237 as a trace [blank]. This is made in a by-product reaction nuclear reactor with an effective fission yield of one-tenth of a percent, and I had assured him that we could get him some if we thought they were using it effectively. We would make it at a sacrifice of time. It happened that this was finally made by me and another young man, Howard Mehn [*sic*], an oddball character who soon left the Project as being a claimed homosexual—no, an admitted homosexual.

So I had my only visit to the laboratories of Y-12, which I passed every day on my way to work. The compartmentation between these projects was total, and as a special service to them I was permitted to go. I was met by Kamen and a security officer, I was brought by a security officer—some official of the Tennessee Eastman Company. We were taken to an empty building just being put into service, a laboratory. I was charged to talk about nothing but uranium isotopes and not to mention methods of production. Then I would talk about how they could use this U-237 that I could magically produce at a certain specific activity on their problems, but they were not to disclose any of their problems to me. A totally sterile contact except for Martin Kamen's brilliance.

JBS: The security officers were present all the time?

Coryell: Sure. That we not violate the rules. That I not indicate how we'd make it and that they not indicate why they wanted it.

I then went back and slaved away and produced some millicuries of U-237 with my own lily white hands and messed the hot lab up because I went out to dinner and a pot bubbled over that I was de-etherating by bubbling air through it, and we had an awful mess to clean up. Whether this thing ever gave any results to Y-12, I never would know. It wasn't in the rules of the game and I'm sure it was a mess.

Kamen shocked me with the contempt with which he treated this man who was roughly a vice president of the Eastman Company. He made me embarrassed, how insulting he was to the man. This is the secret of Kamen's great problem. He was very intellectual and he wouldn't stand for stuffiness, he wouldn't stand for incompetence, and he wouldn't stand for bossiness by a man who had a high position if he didn't have the corresponding knowledge.

Kamen also got in trouble with security officers, I heard from a mutual friend who was in the Army at Y-12 but later came, at the end of the war, to X-10, because Kamen was able to arrange for an irradiation of sodium chloride, to get some radioactive sodium for tracing. He was very keen about the use of radioactivity for ordinary chemical problems and he arranged that a gram of sodium chloride be put in the X-10 reactor for a few hours—I don't know the exact time—and be brought in a lead shield to him with the proper Army boys and secrecy. No one at X-10 knew what it was or why it was to be used, and nobody at Y-12 was to know where it came from. It was to appear as if by magic from the Berkeley cyclotron, a bunch of horse shit. The stuff came out of the reactor so dreadfully radioactive it set the automatic alarms off and caused a local crisis until it was put behind enough lead to protect it. Its delivery had to be delayed while it underwent decay. It has a 12.8-hour half-life.

Kamen, upon receiving it, realized instantly our neutron flux was more than 100 times what he estimated. Our production rate was therefore much more than 100 times, he estimated, and we were going to win the war and not them, and he let out a shout of delight and let some other guys know what the neutron flux was, and he got holy hell.

One of Martin Kamen's closest friends was Waldo E. Cohn, who's been talked about quite a lot here. Cohn had known

Kamen when Cohn was a graduate student in biology at Berkeley and Kamen produced radioactivity for the biologists to use in tracing. Cohn had done 12 or 15 papers on radio-tracing and worked at Massachussetts General Hospital before he came to Oak Ridge. He was one of the few radiochemists with real background. They also played music together. Cohn is a cellist and Kamen a violist I believe, and had lots in common up and down the line, including Jewishness. Both were dreadfully Jewish and proud of it, though Cohn was respectful of, though not believing in, orthodox Jewry, and Kamen was a completely atheistic Jew.

I heard the story from Kamen at the end of the war of the details of his being fired, and I think I'll delay these till next hearing. I'll just tell the impact on us in Oak Ridge who knew him.

Now, Kamen had only visited Oak Ridge a few times at Y-12. He probably was under orders not to visit too much with people outside that. Los Alamos people were under explicit orders not to visit, but this was often violated when they came. But he was up at Waldo Cohn's all the time and I was there every time he was there. He was up at our house the time I tried to titrate him.

Sometime in June 1944, or perhaps May, Waldo Cohn gets a letter from Martin Kamen that Martin's been fired by the Army. Waldo was furious and Waldo came to me and to George Boyd and to the few other people who had met Kamen on visits. Most of the people in Oak Ridge had not been in radiochemistry before the war, except Isadore Perlman, whom I think Waldo would never go to because it was felt that the Seaborg kind of mentality wasn't generous enough to take care of an errant man like Kamen. Waldo wanted all of us to agree to quit the Project at the gross injustice of firing a man of Martin Kamen's ability on a trumped-up security charge. As far as I know, he had just a letter from Martin Kamen that came in the ordinary mail.

I was greatly impressed by Martin Kamen. I knew him moderately well, and I would be very unhappy if any man were fired on a trumped-up security charge, but I couldn't see myself sticking out my neck until I knew a lot more about the case. Waldo assured me that Martin's mail would be opened, and that I should not correspond with him directly, but that if I sent a letter to Waldo Cohn's aunt, who lived next door to Martin Kamen, she would pass the thing to him, safely. I was also contemptuous of this method of corresponding with a friend of mine. If I had the confidence that Kamen was all right, I had the confidence to write to him directly. It was obvious that I would have to avoid classified material, but this was not a piece of classified information. So I wrote Martin Kamen directly, probably to his laboratory address, and Waldo presumably did what he advised me to do, write by his aunt.

It happened that in 1948 Waldo Cohn lost his security clearance in the AEC cleanup and was on the classified list for some months until he was reestablished with a hearing, and one of the charges against him was that he'd corresponded with a person known to have been fired. I, of course, corresponded with the same person, but the FBI wasn't concerned about my mail since it didn't come indirectly and the FBI apparently did enough snooping in the mail to know that Waldo was corresponding by this presumably safe method, but the fact that he was using this method indicated he had some concern about the situation which might imply he was equally guilty of whatever they thought Kamen was guilty of. This has always been a principle of mine. If I don't like what goes on, I'll not use hidden methods of communication and so, wrong or right, my position on most of these things is public knowledge with carbon copies everywhere and shouting in the halls, and things like this, and I never have to worry about being charged with secretly communicating with anybody.

The tragedy of the situation is the way Martin Kamen was treated. I got several letters from him during the war and they may be lying around my files somewhere. Martin wrote me a good deal, and wrote Waldo Cohn bitter letters about the personal treatment, about the way people from under the rocks—it appears that lots of people he thought were friends were people under the rocks. [Donald] Cooksey (1892-1977), who I don't know, was a man he was particularly bitter about, because Cooksey had been an important Project officer and had always acted like a friend of his. E. O. Lawrence himself was out of town when things happened, and apparently was very much distressed. By the time he got back things looked irreversible. Kamen had unburdened his feelings on some more Army people.

I might say that the technicality he was fired on was having dinner with two Russian consular officials, who invited him to dinner. He thought it would be a good fishing expedition and he could fish as well as they, and in the process of firing him for what he was purported to have told them—an Army security officer had been·sitting at a nearby table taking notes, the man having no more scientific knowledge than the one in the bus who thought illinium was uranium in my case, it was a massively distorted body of science. But the Army security officer never asked Martin what information he'd gotten from the Russians, and he'd been speaking a good deal about what they knew about Oak Ridge. This information has never been tapped by the government to this very day.

Most of the people he thought were friends cut him cold and were afraid to death of him. The Army said he couldn't stay at Berkeley because of policy security has against having working near his old haunts. There also was, as I may have indicated earlier, the general feeling that security was loosest at Berkeley because atomic energy was oldest there in the chemical field and the radioactivity [research] was oldest in the whole United States.

James Franck thought that Martin Kamen would be a wonderful person to put on his photosynthesis project which he still ran as a sideline under Hans [? Seuss, 1909-1993], and the Army security officers refused to let Martin Kamen work in any city where atomic energy was prominently being worked on. The last minute they relented and said he could stay at Berkeley in any non-classified job he could get, so he got a job at the Kaiser ship-building concern, at night, titrating for sodium chloride in the bilge water of new ships, a perfectly routine chemical operation that a high school girl could do. And daytime he worked with carbon-14 that he'd made himself before the war. He discovered carbon-14 with [Sam] Ruben. And he was able to do a reasonable amount of chemical research under difficult circumstances.

However, in 1945, before the war ended, Arthur Holly Compton was appointed chancellor of the Washington University of St. Louis and the first man he hired was Martin Kamen, who moved to Washington University before the end of the war. Compton just told security he wouldn't stand for this nonsense. He had reason to believe that Kamen was totally honest and very able, and he apparently felt badly at the stories he'd heard of Kamen's treatment. He didn't know Kamen personally, but when Kamen was attacked by the House Un-American Affairs [Activities] Committee in 1948, the scientist X at the Washington University of St. Louis, I was in Boston with some doctors' examinations, so I can date the thing September 2nd, plus or minus one. Grace Mary and Julie were down at Brookhaven. Soon as I saw the name scientist X, I knew it was Martin Kamen. I wired Kamen that I would give him full support within the limit of my qualifications.

I was delighted to find in the evening issue of the paper, that Arthur Holly Compton identified scientist X as Martin Kamen and gave him his full support. Kamen went to the hearings and fought hard for open hearings before the House Un-American Activities Committee, and failed. But his hearings were later published by the House Un-American Activities Committee with their editing and without his.

JBS: What were the grounds on which the [HUAC] Un-American Activities—

Coryell: He gave classified information to the Russian consulate officials at this dinner.

JBS: I was wondering what were the grounds of the Un-American Activities Committee maintaining a closed hearing?

Coryell: It was standard practice. You had a tough witness, he'd make a fool out of you.

Kamen was willing to talk frankly but bluntly. Even after the hearings were edited by them, without Martin Kamen, if you read these hearings critically you'll see that he made a fool out of this committee at all times. They didn't know what they were doing, they got no information, and they didn't know what the score was. Martin Kamen knew what the score was and was willing to talk frankly, but was very uncooperative to letting them press him into their image of him and what they would like to make of him to the world.

In this same set of hearings there are two other people being interrogated. One of these is Clarence [Frances] Hiskey, whom I knew slightly and liked little, who was apparently involved very messily in spy information and was not willing to talk at all. The House on Un-American Activities Committee got no information out of him, nor did they ever bring the situation to a point where they would indict him. There was a third man whose name slips my mind now, whom I'd only met once or twice. He was another Columbia man who'd been in Chicago a short while, who apparently did a great deal of soul searching, talked guardedly but completely, and one could never tell what really went on, was he crooked or not, or just sort of playing footsie. But the House Un-American Activities Committee also was unable to establish any case against him and was able to get no information. These three cases are total failures of this committee interrogating in private session, with the information from their lawyers and their research group, to make any contribution to the question of has subversion involved science or were these three scientists involved in any criminal or subversive activities. Kamen was not, but the mass of the evidence was against him because of the bias of this incompetent recorder who took information—reams of it—that has never been fully evaluated yet. But it's been 95% evaluated by the FBI for the purpose—I'll come back to Kamen again.

Session Five — 14 August 1960

Coryell: It's just a month ago today we started this.

JBS: That's right.

Coryell: A lot of water has gone under the bridge. Each one of these meetings I've been just as rushed getting to, and just as rushed getting away.

A good deal more could be said about Kamen and it may very well be that the Oral History Department would like Kamen's own story. He is presently professor of biochemistry at Brandeis University in Waltham, and it's my private opinion that he's a very good candidate for the Nobel Prize for his discovery of C-14 with Sam Ruben. Sam was killed during the war in a phosgene accident at Berkeley. Sam was the man I mentioned in my second talk that warned me about Seaborg, and I tried very hard during the war to get Sam to join the plutonium project, specifically with the aid of James Franck. I made every effort I could by telephone to persuade Sam Ruben to take my position at Chicago when I left, to have a strong, able radiochemist and I would hold the front at Oak Ridge. Sam was unwilling to leave Berkeley. He felt he owed a lot to Latimer. I believe, on the basis of information from people like Kamen, who knew him well, he felt that a Jew had less than average chance in a university and, if Latimer leaned on him heavily, he would never let Latimer down. Latimer never seemed to give much attention to the Manhattan Project, except he was nominally in charge of the work at Berkeley and he visited Chicago every three or four months. I think he visited Oak Ridge only once.

I don't know whether I am talking about Kamen, Ruben, or Latimer. I think I'll finish Latimer. I knew Wendell M. Latimer, professor of chemistry at the University of California, before the war. He was an eminent physical chemist, a vigorous man, a hard man, and a cynical man. It was always a pleasure to talk with him. He had good sense about chemistry and didn't stand for nonsense. So I had a friendship with him predating the war and I was therefore very inclined towards him when his name was discussed in the summer of 1942 for the job at Chicago. It was only when I realized that he didn't give a damn about the Seaborg case—he felt that Seaborg was able and could take care of himself and would not concern himself about people I thought Seaborg was hurting— that I realized that Latimer would be a careless director, not necessarily a tool of Seaborg, but that he would just be an absentee landlord. He had another empire at California in the gas warfare work, where Sam Ruben was one of his chief men.

I knew Latimer well enough so that when I saw him when he visited Chicago, approximately February, 1943, I challenged him for letting Seaborg run wild on the Project and for the damage Seaborg had done personally to Joseph W. Kennedy and Charles D. Coryell. Latimer dropped the cigarette he always carried loose in his mouth another 15 degrees, and said, "Seaborg made a man out of a mouse in both these cases." I felt completely deflated. Seaborg had made me a far better chemist in trying to match him scientifically in order to fight with him politically, and Latimer was essentially right.

Kennedy had had his problems with Seaborg when Kennedy was a doctoral candidate under Seaborg. There was a standing joke about Kennedy having Seaborgitis. He was ill, was thought to have tularemia—I believe there was an exploratory operation to see whether there was anything wrong internally—and all these boys in the laboratory claimed he was suffering from Seaborgitis. They ascribed my 35-pound loss of weight during the first four months of the Project to Seaborgitis, too. But it is a disease one recovers from.

I saw Latimer several times in Chicago, and I liked him. I saw him in Oak Ridge on his only visit down there, and I suppose I saw him in 1945, after the war, when I visited Berkeley when I was still planning to go to UCLA. I have always been interested in his position in the Oppenheimer case. The testimony in this nine hundred and ninety some-page case is a rather interesting commentary on human feelings, and I think the most idiotic testimony of any man I respect is that of Wendell M. Latimer. I can cite the person who wants to check into this to Latimer's own words. But most of my friends cannot understand that Latimer was an intelligent, hard-thinking man. My interpretation of Latimer's position is he was utterly contemptuous of Oppenheimer, all this idealism about moral principles about hydrogen bombs, he was jealous of Oppenheimer's empire, he probably had had differences with Oppenheimer about the staffing of the Los Alamos project because he was competing with Seaborg for chemists and it turns out that the people who went to Los Alamos were those who wouldn't go with Seaborg to Chicago because they didn't trust him and Los Alamos robbed these heavily and Latimer always wanted a branch of the plutonium project in Berkeley for the prestige of Berkeley. My opinion is he went into the hearing room before the Gray committee [AEC personnel security board: Chairman, Gordon Gray with Ward V. Evans and Thomas A. Morgan], all the testimony was sworn to secrecy, a man had to promise not to speak outside, so he had total coverage for secrecy. Oppenheimer was in the room, so whatever was said Oppenheimer knew what was going on and Oppenheimer could figure it out for himself, and he decided there's no point in talking to the Gray Committee about the grays of science—it's either going to be all black or all white. He decided for the benefit of God, country, and the University of California, he would make Oppenheimer as evil-looking as possible and the things he stood for as bright-looking as possible. If I read his testimony in this framework, I realize that Latimer, a cynical man, was trying to convince the committee that Oppenheimer was a bother, a headache, and a bad influence on the Project, and it's too bad if he gets massacred in the deal.

Now, the Gray Committee material was published without warning by the Atomic Energy Commission by [Lewis Lichtenstein] Strauss (1896-1974) [pronounced "straws"], on the technicality that [Eugene M.] Zuckert (1911-2000) had lost it for a few hours and it might have been read by newspapermen and therefore must be considered as probably public knowledge. It was hastily declassified, and

192

no person testifying had a chance to modify his testimony, but Strauss' men could do what they wanted with the government's side of the testimony. I don't know how much the testimony suffered by the editing.

The security editing is very, very poor. It says, in one place, "the Project district consisted of three parts." Many pages later, the three parts are separately identified by first, so-and-so, and then pages later, the second part is so-and-so. I feel definitely that there are security slips in the Oppenheimer testimony, if you judge security by the rules that Strauss expected other people to follow. But then, politics always overrides science, and Army security.

Latimer was in the mountains of California in the early summer of 1955 with friends and suffered a heart attack that wasn't recognized as a heart attack, and died.

JBS: Do you know of any response Latimer had to the fact that the testimony was made public?

Coryell: I'm sure that Latimer felt extremely embarrassed, because he really made an ass of himself, but I suspect he told people who came along, "Well, by god, I had to do something to try to straighten this thing out, and I decided to say this. It may look silly, but...."

Well, I shouldn't try to put words in his mouth. People in Berkeley know what he said. The only university center where Oppenheimer was under steady attack was the University of California, and virtually everybody who testified was from there. Going further into the Oppenheimer case, it's also interesting that Seaborg did not testify. Seaborg was the only member of the [AEC] General Advisory Commission who felt obligated to go ahead on the crash program for a hydrogen bomb. He was going to Europe—I think this was in connection with his Nobel Prize [1951]—and he had left a letter for Oppenheimer saying that he was reluctant to do this, but he felt we must go ahead. And Oppenheimer consciously or unconsciously suppressed this information, and a fair amount of testimony involved to what extent had Seaborg's position been distorted consciously or unconsciously by Oppenheimer. Seaborg told me some about this, oh, in 1956, and he said that neither side trusted him to testify and that he was glad he didn't have to. He felt that this was an unholy mess and largely a personal conflict.

Now I should go on with Martin Kamen. When I first saw Kamen after the war in 1947, when we were both speakers at a conference on use of radioisotopes in biology and medicine held at Wisconsin (the results of this conference came out as a book), I was just delighted to see him. But already there was a security buildup within the Project and dozens of friends of mine had been in security trouble, and I felt it appropriate to have with me an arbitrary third party as an observer when I had my first long talk with Kamen. The man I selected, by accident, was Richard L. Kenyon, who's now editor of *Chemical Engineering News*. He was a reporter for *Chemical Engineering News* at that time, and I think it was extra nice to have a reporter hear the story. I'll enlarge slightly on why I had to make this decision.

Kamen told me at great length the story of his treatment at Berkeley and, more important, he told me in great detail his relations with the two Russian consuls. I don't think it's appropriate that I repeat the Kamen story in more than say a high spot, because Kamen is available for this story and the story is quite interesting [Kamen 1985]. The House Un-American Activities Committee didn't find it. Let's just say this, that a Russian consular official at some general San Francisco tea involving aid to our allies, approached Kamen to find out how to get phosphorus-32, which was and still is used as a treatment for leukemia, since a consular official—I believe in Seattle—had leukemia. Kamen referred the man to John [H.] Lawrence (1904-1991), a brother of E. O. Lawrence, who was the medical director of the radiation lab. Radioactive phosphorus was provided and the Russian consular people were very satisfied, and I believe they made a gift of a case of whiskey to John Lawrence, as I recall Kamen's testimony.

Kamen they invited to dinner. Kamen was very interested in this fishing party, but the great mistake Kamen made was he did not invite with him any arbitrary American person, any person, who could testify what the temper of this meeting was. Kamen said also that the Russian people had approached him to get instrument catalogs for the Russian purchasing agency because the catalogs were in short supply during the war. There were so few being printed, and so many government agencies were buying everything in sight. So Kamen located a large number of catalogs of various companies, including particularly Leeds and Northrop, whose catalog had a certain number, such as K-37. Kamen brought this stuff to the dinner. He was seen by the counterintelligence observer passing documents over and there were numbers on these, and some of these numbers [blank] testimony.

There was about an hour-and-a-half wait at Fishbein's Sea Grotto [Bernstein's Fish Grotto] to get in, and they were drinking and talking and waiting in line, and, as I say, there was a long fishing party where the Russians were trying to find out if Kamen knew anything about Oak Ridge, and Kamen was trying to find out why the Russians didn't collaborate with some fancy artillery stuff he knew about that we found out about by capturing some material the Germans captured from the Russians, and the ballistics involved would have been of great interest to the Allies. Kamen had picked up information like this in order to press the Russians on collaboration. The Russians said Kamen would certainly be welcome in Russia after the war, indeed during the war, and Kamen was hoping to get a free trip to Russia out of this deal. It was the next day the security officers called him up and summarily fired him for talking with the Russians.

Now, later on, in about 1948, at the time the House Un-American Activities Committee discovered this situation and called Kamen scientist *X*, the *Chicago Tribune* published a picture of Martin Kamen giving documents to the Russians. This picture is surely a faked picture. Kamen says that the man who faked this picture was a protégé of Strauss and was later hired by Strauss and protected by Strauss. Kamen sought support in various places for testifying, before he went before the House Un-American Activities Committee, and some people who Kamen thought highly of were unwilling to take the risk of giving character testimony for Kamen. I'd already wired my willingness to, but I wasn't used since I wasn't at Berkeley these days.

But there were some men who'd shown no interest at all in politics, as Kamen or other people knew, who were willing to speak on the basis of their long friendship with Kamen. Since Kamen was a very frank, open character, I felt that I knew Kamen pretty well on the basis of two or three pre-war contacts and two or three wartime contacts. One of these men who offered to testify freely was George E. Boyd, of the Oak Ridge National Laboratory. Boyd, in the atmosphere that existed about the Kamen situation, was risking his job for sure. He's now assistant director of the [Oak Ridge] laboratories; he was then a group leader. This was particularly true if Kamen's claim that Strauss' protégé was the person involved in falsifying the information.

Now I'll close this discussion about Kamen by saying that Kamen brought a suit against the Washington *Herald,* then a subsidiary of the *Chicago Tribune* in Washington—a libel suit. Nathan David was his lawyer. And eventually he got a settlement out of court from the *Chicago Tribune.* Practically nobody wins libel suits against the *Chicago Tribune.* There's one more aspect of the Kamen story which, if the tape lasts long enough, I'll sometime talk about. This is the problem of passports.

Kamen was invited to Paris, to the symposium on Isotope Exchange and Molecular Structure, which I was invited to. My way was paid by the Rockefeller Foundation but my stay in Paris was subsidized by the French government. [W. H. Burgus, T. H. Davies, R. R. Edwards, H. Gest, C. W. Stanley, R. R. Williams, et C. D. Coryell, "Étude des États de Valence des Produits Formés Par Désintégration Bêta et Transition Isomérique," *Jour. de Chim. Phys.*, 45:165-176, 1948.] Madame [Irène] Joliot-Curie (1897-1956) was the chairman of the organization. Kamen was unable to go because he was denied a passport, and the passport denial I'll tell about later. I forgot—I thought that this preceded the passport denial.

I was M. le Président [presiding over a session?] when M[oïse] Haïssinsky (1898-1976) read a French translation of Kamen's speech. During my attendance at the meeting I was taken to Joliot's house and heard [Frédéric] Joliot (1900-1958) describe Mrs. Joliot's experience being put into the jug at Ellis Island on arrival in the United States in January, 1947. She was making a trip on behalf of Spanish refugees. She had a visa properly received, but the Attorney General's office imprisoned her on arrival and held her overnight. A ground wave of telegrams from all over the country to President Truman led to his action to release her the first thing next morning and made her trip an extraordinary success. She raised four or five times the amount of money she expected, she was satisfied that her feeling that the American government was idiotic was well proved, and she found that she had many more friends in America, so in the long run she wasn't hurt.

Joliot, in these discussions, and in other contacts I had with him, was very pro-American. He liked the way Americans did things. He told me, this particular evening when I went to talk about the planning of when my talk would come and what would happen in Kamen's talk, he said, "After all, America's a democracy. The only requirement is there not be more than 50% idiots in the government." And he said he didn't think the government was that idiotic.

I had the strong feeling that the Kamen case would involve guilt by association between him and Joliot and for that reason I was particular to indicate that I would support Kamen because I knew what went on with respect to—I knew, when he was going to receive a French invitation, I told Haïssinsky he couldn't get a passport. Then I was the official at the morning when his report was given, and I know a good deal about the Joliot relations with the conference. I have the feeling that my relations at this conference were utterly kosher, though Joliot was, of course, a member of the Communist Party at this time. Madame Joliot had had visa difficulties and Haïssinsky was denied a visa for many years and eventually had a visa taken away from him in Boston that I was involved with, but he is again in the United States. Apparently he can come in through the Attorney General's office. I felt that Kamen might have a very difficult time answering questions of the House Un-American Activities Committee about his good friend Joliot when he'd never met Joliot.

I'll say a few more words about Kamen's passport. Martin Kamen apparently—this must have been just before the French invitation and I can't date it exactly now—was invited to go to the Weizmann Institute of Science [Rehovot, Israel]. They wanted him to head the biology division or the biochemistry division of this institute. At this time I knew very little about the Weizmann Institute, though I later was a guest professor there [January-June 1954]. He called on Louis [Frederick] Fieser (1899-1977) in Boston, since Fieser was a great American friend of the Weizmann Institute, and talked to me over the phone. [Kamen] had left his passport in New York at the travel agency the Weizmann Institute uses to get his Israeli visa and the other parts of his trip straightened out, and to pay for the ticket. A representative of the Attorney General's office came into the travel agency, flashed a badge, and took away Kamen's passport.

Kamen immediately wired Mrs. [Ruth B.] Shipley [chief of Passport Division, U. S. Department of State]: "My visa has been stolen." Mrs. Shipley wired back: "I confirm the action of the Attorney General's office that your visa has been removed for proper reason."

Kamen checked, and went down to see her, and he says she was hopping mad that the Attorney General's office had overridden her. She had given him a passport knowing about his record in the Manhattan Project, and when she gave a passport she was sure a man was an honest American and it made her very, very angry she was overridden. On the other hand, her hands were tied and all she could do was urge him to go see the FBI.

So he went back to St. Louis and called on the FBI, which was delighted to see him. He went in the morning and they worked all day—they took him out to lunch, they took him out to supper, and they were able to explain 95% of the testimony in this many, many pages of handwritten material this counterintelligence agent had picked up in Morris Fishbein [San Francisco restaurant, Bernstein's Fish Grotto, spring-summer of 1943]. They wanted, if possible, to find a Russian spy to close the case absolutely, though they did expect he'd explain the remaining 5% of gibberish before they gave him a total clearance. But they were very warm to him because he helped clear up a big puzzle. Kamen and I both feel that the FBI is very angry at the Atomic Energy

Commission because of the way it has handled counter-intelligence files.

As near as I can tell, when the Manhattan Project was turned over to civilian control, January 1, 1947, people who were in good standing in the Project had their files just say, "Case okay." The huge file which exists on me from the fights I was on in the Manhattan Project is either in AEC files or destroyed. The people like Kamen, who had been fired, had their files turned over. The Atomic Energy Commission, having fired Kamen, had no reason to evaluate these files, and the House Un-American Activities Committee made a hopeless mess, and the FBI was puzzled though they would have been anxious to clear it up.

The FBI was particularly kind to Kamen, and so was Mrs. Shipley, because the week this happened he was listed in *Time* Magazine as an eminent biochemist at the Washington University of St. Louis for some work he did, I think, with a man named Perlman [Isadore] on nuclear protium [Kamen 1947].

Kamen has finally gotten his passport and traveled, whenever he did, with pleasure, and apparently his status is all right but this came after his hearings before the House Un-American Activities Committee, which I suspect was the spring of 1949—I don't have the date just at hand. Kamen, I guess, had to sue for a passport about 1953. He sued for a passport after he'd won a libel suit against the *Washington Herald* and a libel suit against the *Chicago Tribune*.

JBS: And the passport was granted at that time?

Coryell: The passport was granted as a result of a suit being instituted that the State Department didn't want to defend itself against. All this caused Kamen lots of money and trouble, but he was glad to fight. His prize possession is the check he got from the House Un-American Activities Committee. He had the stub, I mean the receipt portion, framed in his study: "For Un-American Activities, $30.00," or whatever the amount was.

JBS: I think we might back ourselves up.

Coryell: I'm thinking of other problems that involve political tension and fight with the Army that happened in the Oak Ridge wartime years. One thing that always gave me a great deal of delight was the time I thought I had made a brilliant discovery in scientific counterespionage.

The date of this must be about November 1944, and there is, in classified files, a memo or more from me on the subject. Within several months after D-Day, an article appeared in a Canadian newspaper by one of the well-known United States commentators—I think he was United States, the article is worth identifying—which described a very curious installation which was found when the Canadian troops took the Pas de Calais. It appears this same newspaper article appeared in many United States papers, but most cases—though not all—known to me had one paragraph removed which was the most descriptive. The newspaperman saw with his own eyes a large cement building with walls six to ten feet thick, with all the plumbing outside, with thick doors of steel, with three large rooms roughly 50 feet on a side, and he couldn't imagine what these factories were built for. One of the soldiers working with me, Gerald Strickland, saw lying in the GI USO [government issue, United Service Organizations] recreation room at the barracks this clipping. Some Canadian mother or relative had sent it to her son in the United States Army in Oak Ridge. I guess there were Canadian boys in the United States Army in Oak Ridge— these were all special, cleared personnel, and Canadians wouldn't have been excluded.

I recognized the situation instantly as a hot laboratory. The only hot laboratory in the entire world known to me up to that time was the 706 C building I described to some extent earlier, except that I was able to visualize right away a reason for having the hot laboratory. The Germans, if they wanted to do radiochemical warfare, they would want efficient reactors, make lots of fission products, load these into V-bombs, and fire them out. But for preparing the fission products you need lots of shielding, you need outside plumbing for maintenance, and the big room inside would be the reactor room. I was thinking in terms of a heavy water reactor with shielding such that this would be the cheapest way to make fission products. I also knew from this small amount of rumors we got from England, that for some years there had been a squadron in England that searched for radioactivity in all bomb craters. You see, [perhaps] the Germans were playing with radiological warfare. It was [would be] the first time it was used in quantities because of considerable biological damage. But above all, it would be a wonderful terror weapon.

I also think I may have mentioned that I saw in *Time* Magazine pictures of big guns on the French coast with the name "Urania" in November, 1943. So it all looked as if

Hitler might have tried radiological warfare and be prepared to go ahead on a big scale, and the invasion of the French coast removed this. It's also obviously what these plants near the V-2 rocket bases were, which were all around the Pas de Calais.

Well, this alarmed me very much and, with my ability to rationalize quickly on no evidence, I decided that the only answer to this was some chemist or physicist who knew hot laboratory operation and fission should be sent instantly to this territory which was now in Allied hands. I talked this over with Warren Johnson, the director of the chemistry division at Oak Ridge, and he agreed there was a good chance I was right. He took me to see Doan, or I saw Doan, and Doan thought that my ideas were very reasonable. Warren Johnson agreed to talk the matter over with Hogness, his good friend in Chicago who was then director of the chemistry division there, and Johnson and I went to see Arthur Holly Compton, who was leaving for Washington.

I prepared a memo, which had to be highly classified, for the benefit of the people involved and I saw that copies went to friends of mine in other projects who knew about hot laboratories. I think one of these was probably William H Sullivan (?-1966), who was then at Hanford. I don't know that I sent any to Los Alamos, but I may have sent one to Dodson—Dodson will know if I did. I sent one to Sugarman, as I sent copies of everything like this, and I think I may have talked guardedly on the telephone to Sugarman after he received this. But I thought I had the hot stuff.

Compton gave us assurance that he would see this went to the proper authorities in Washington, and some days later he said it went to the highest position. Several times during the discussions I was asked why I didn't take this up with Army security, and I expressed my utter contempt for Army security, that they were idiots, and they would block it up with red tape, and they would make it so secret nothing was done. Approximately two months later I was asked to see the Army officer at Oak Ridge, Edward J. Murphy, who is a man I like very much.

He told me with pain, very much like the pain he had when he expected to fire me in the story I told two times ago, that I had done this in totally the wrong way and I had made considerable trouble for him. The proper way, when you find a security break, is to take it to Army security. And I expressed my contempt for Army security. Then he tried to remind me, on top of this, this came to me from a soldier and the soldiers are committed to Army security and had no choice, and I had made this thing very difficult for the soldier. I said, "I don't give a damn. I have my duty, as an American citizen to do everything I can to protect the security of the United States and, if I have to go up to the President of the United States and get fired for it and have ten other people fired for it, if I think I'm justified in a security situation, I'll do this. I can't help it if it hurts your feelings. If you come and ask me what I've done last week, I'll tell you. But if you try to block what I do when I do it advisedly to avoid your office, I'll also do anything I can to break out of the security department."

The story now seems to be—and no feedback ever came to me—these were hydrogen peroxide plants. The Germans had developed during the war the ability to handle, to make and handle, 95% hydrogen peroxide, well in advance of the technology in our country at the time. It was part of the rocket fuels for the V-1 and V-2 rockets. Here again, you have outside plumbing because of the explosion danger inside. And you had thick walls to contain the frequent explosions they had with imperfect technology. And you had it close to the French coast because to send the V-2s off you don't send them by railroad car. But, from the description given by the newspaperman who obviously had good observation abilities, one couldn't tell whether this was explosion-dangerous material or radioactive-explosion-dangerous material. An instant look at this thing would have told us, and had it been radiological warfare, all my thoughts would have been more than justified. So I felt like a small hero anyway.

JBS: This likely went onto your security files, too, that you hadn't told the Army.

Coryell: It may have gone to my credit in the security files. I think that any security officer would have realized that I did a thing he couldn't have done. I don't worry about this. But I'd love to see my file.

Matter of fact, Sam Allison once mentioned he'd seen my file and it was an inch thick. But this may have been a total joke—Allison loved me very much and liked to tease. On the other hand, the file that the FBI has on me was apparently very small in 1948. In 1948 the House Un-American Activities Committee was raising hell about atomic scientists and the Federation of American Scientists had a fairly good inside track about what was going on—as a matter of fact, anybody who read the Hearst papers could tell the next move the Committee [would make] because the Committee [blank], and the story went around that there were 15 scientists on the list and I deduced I was about position 13. Had I been Jewish I would have been in the first five positions, or had I not been at MIT. But MIT looks so conservative this gives me some protection, and not being Jewish gives me some protection. I was somewhat disappointed when I was never tripped up, because I'd have loved to have had an open brush with one of these committees on the basis of the fact that the charges they were making about some people, for which there was no justification.

I think I should also put on record the statement that to the best of my knowledge only one United States scientist has ever been indicted for espionage. This man is a biochemist named Harry Gold (1910-1972), who was not on the Manhattan Project, but was involved apparently in the Rosenberg case. [Martin] Sobell and [Julius and Ethel Greenglass] Rosenberg and the other person there, [David] Greenglass, were not scientists. They may have had some technical education but they were soldiers in a technical capacity, and their security violation has nothing to do with American science.

Among the scientists that have been proved spies are the English citizen [Klaus] Fuchs, a German refugee with a Communist background who'd been in a concentration camp; Allan [Nunn] May (1911-2003), a very English Englishman; and Bruno Pontecorvo (1913-1993), a volatile Latin [Italian], who was a British citizen, as far as I know. I met [blank] [not Fuchs] briefly; I met Allan N. May a few times when he visited Chicago for a six-week period in 1944. It's also interesting to point out that Allan N. May was put in charge of the experimental physics group in Montreal in the Anglo-French Canadian project because John [Douglas] Cockroft (1897-1967) and the higher people wanted a good king's man—not have a bunch of unreliable French Russians in charge of the project.

Allan N. May's security defection, as nearly as I can establish from all I've read and what I've talked about with French Canadians and English, was largely an independent rebellion in the interests of world understanding as he saw it. He thought we had to make some gesture to the Russians to have any chance of an exchange later. But it's quite clear that the Russian espionage system was glad to receive his information and did everything it could to compromise him [*sic*], which involved gifts of some whiskey and a small amount of money. It surprised me when I went to France and England in 1948. I was with a bunch of people who knew him well, I made my second contact with members of the Anglo-French Canadian team. Everybody I talked to was sure that May was guilty and should have gotten a sentence, but the people I remember talking to were also shocked that he got the maximum sentence rather than the minimum sentence. But it is true that he gave U-233 and U-235 in 50- or 100-milligram quantities to the Russian espionage system, and it got to Russia.

I might also point out that this spy system that did involve him in Canada, involved for the rest, Canadian scientists who were in high explosives. The discovery was made by the defection of the Russian official, [Igor] Gouzenko (1919-1982), not by technical espionage. As far as I know, the only spy discovered by technical espionage was Fuchs. The story went around the Manhattan Project and the Atomic Energy Commission laboratories in this country that in 1948 and 1949 the rumor level in English physics about hydrogen bombs and fission bombs was high, as American observers of classified nature found out. This only could have come from a leak at Los Alamos, and the leak could only involve one of seven or eight people, because there weren't that many more Englishmen at Los Alamos that could have had this leak. Fuchs was caught by starting with this observation. This was the first case, and as far as I know, the only case of scientific espionage.

On the other hand, the American public felt, in the late forties, that there were many idiots in science and many dishonest men and the government science labs involving defense work were riddled with espionage. This was by no means true. I think there were more people involved than Gold who were scientists and who did things dishonest with security, but apparently none of these cases was well enough established for an indictment.

The one of which I know nothing about the details, except for what I read of the government hearings, [the Hiskey] case, may come next to this. The Army at least felt, when he was at Chicago—he was a reserve officer—and to get him out of the situation, they assigned him to the Alcan Project [near Mineral Wells, Alaska] to get him geographically away from people. He served his stint over there. He was called a spy by the House Un-American Activities Committee—he had a hearing I've already mentioned. He says that the head of the House Un-American Activities Committee at this time—I think it was probably Congressman Wood—was horrified when he walked into the court with his arm around his present wife and his arm around his ex-wife, because they'd counted on playing husband against ex-wife.

Also, in this same connection, in 1948, I expected to be one of those called—I don't know, I thought because of my Joliot connections, you see, of which I was very proud—the Joliot connections I've told you about. I must add the fact that Madame Joliot was making this tour of the United States and some organization in Boston was sponsoring a dinner, and I was asked by postcard would I be a sponsor of this dinner. Well, of course I felt obligated to be a sponsor of this dinner since I was to be a few weeks later a guest in Madame Joliot's laboratory. So I'm on the letterhead as a sponsor of this dinner—I'm very proud of this. Since I was in Europe I wasn't present at the dinner, but several of my students attended. Madame Joliot gave a talk in English and she was somewhat hard to understand, but she was very bitter about the United States. But the people who went were very glad to see a very, very great chemist.

I had three or four discussions about American spy problems and American Un-American Activities problems with Madame Joliot, and she was extremely bitter about their behavior with respect to Spanish War refugees. She— this trip was to raise money for refugees in camps in southern France (these camps still exist). She named some names I don't remember, of people who were fund-raising in the United States, and they were taken by the House Un-American Activities Committee and asked to testify about their sources of information in Madrid. They gave names, reluctantly, with the promise that these names would never be passed outside the Committee. The whole list appeared in the Spanish Embassy a few days later. People's lives were threatened by this situation.

She said this was a definitely known fact. This corroborates a number of other things I've heard about the carelessness with names. The names involved could be established easily. There was also a steady rumor that [Senator Patrick A.] McCarran [D-Nevada] (1876-1954) had a pipeline to the Spanish Embassy, and I've always felt that McCarran was a more dangerous member of the United States Senate than [Joseph R.] McCarthy (1908-1957), because McCarran was quiet and effective and McCarthy was noisy and relatively ineffective. I was very glad when McCarran died, but I think he's still waiting at the gates of heaven for a visa while he's being looked over by the Un-Heaven Affairs Committee. [CDC persistently refers to HUAC as Un-American Affairs Committee.]

I guess I'll talk a little while about the feeling of the scientists as the war came to a conclusion about how the atomic bomb should first be used. This is the situation I discussed with Mrs. Smith [Smith 1958], and I believe my discussion with her postdates her article in the *Bulletin of Atomic Scientists*. The exact date I can add when I get at the notes, but I guess someone told her that I knew a good deal about the Oak Ridge end of the thing.

Oak Ridge was a stepchild of Chicago and Oak Ridge was totally ignored by Los Alamos in the early days of the Project. We just were sort of doing a dirty job, we were sort of finishing up, and it wasn't very exciting. We had very poor internal connections at Oak Ridge, zero connections from one big project to the other. And there were relatively poor connections with the various aspects of the same project because people were busy and there was a sort of tradition of not talking to your neighbors. So I would go to Chicago roughly once a month, and find out a lot more about what was happening at Oak Ridge than I found out at Oak Ridge. I would always transmit information. I tried to develop a technique of talking all the time, bringing information about all these things, with due respect for the security barriers, at the same time having a wire recorder running in my head and carrying back all the information I could from Chicago. The only time I never passed information was if I got, adventitiously, information from Los Alamos. I systematically suppressed telling this to anybody except in one or two cases, where I got strong hints from Sam Allison. I'd see him, and he'd shut the door, and he'd say, "I think you ought to know this." He told me sometime in the fall of 1943 that there would probably be a massive demand for radioactivity on the hundred-curie scale and I should know this because the request might come to me. He was giving advance warning for the barium job, though I think he specifically mentioned cerium and there were reasons why cerium might have been attractive to Los Alamos. Cerium-144. This information I relayed, but he already had the hot lab authorized and construction about ready to begin when this information came to me.

I got some information which could lead me to estimate the timing of the first use of the bomb, and by compartmenting this and not talking about it I also didn't think about it, and I was completely taken by surprise by the details of dating. Most of us knew the date it was going to be fired by the rumors of the bets going on in the Army building. There was a pool on when it would be, with bets from about July 29 to about August 10. The bets involved known production times, with guesses about delivery time—you may remember that the cruiser USS *Indianapolis* that carried the second bomb over was destroyed by the Japanese just a day or two later—plus uncertainty about the weather, which I think did delay the bomb a bit. I just shut my ears to these things. I figured that the time and place of using the atomic bomb were tactical secrets of the highest importance and they had no bearing on the fission product work I was involved with.

I made only a satisfactory guess about the use of barium-140, I think, about the time the bomb was first used. My boys always asked, "What do they want this for?" and I said, "I don't know, and if I did know I wouldn't tell you." I felt that was the way it should remain. (End of tape.)

Coryell: It was obvious to me that the radioactive barium, radioactive lanthanum job was of tremendous importance to Los Alamos, and I think I mentioned I got telephone calls from Oppenheimer daily for about a 10-day period. He stopped us in the middle of a run, and I had to send a telegram to him when the first day's work was over, because he had to know for sure we'd started.

I knew that the timing of the job and the importance of the Project were immense, and I literally feel I was within a short time of being a bottleneck of the atomic bomb. Consequently, I didn't talk about this one until the end of the war. I think I mentioned we celebrated September 18th as barium day at MIT, but this was a celebration for the Oak Ridgers who were with me.

I knew that the Los Alamos code for barium was RaLa for radioactive lanthanum, and this the daughter they had to extract there, and I knew some of the difficulties that Thompson's [?] [Dodson's ?] men had, and later Gerhart Friedlander's, in extracting it. When I visited Los Alamos in 1951, there was a brand new building for RaLa. This building was finished while I was there and was subject to open house for the community of Los Alamos [including] the wives, so I knew I was free to talk about it. But it's rather interesting to point out that in Arthur Holly Compton's autobiography [Compton, A. H. 1956] on his life in the Manhattan Project he chose to tell the story of radioactive barium and called it radioactive strontium. I think this is not an error on his part—I think this is to avoid any fear of security. Strontium is obviously in the public eye. Strontium-90 for fallout and the like, and people talk about it and he obviously thought there'd be no security problem at all to misname it slightly. But it irritated me that he should feel, in 1956 when his book came out, that he couldn't talk frankly about technical matters of the Project. I take it there's no classification about barium-140 or RaLa.

I was talking about how I was one of the information carriers between Chicago and Oak Ridge, and I specialized in talking all the time and listening all the time. People are still surprised that I can dominate a technical conversation, and they think I don't pay any attention to what they're saying, and then I can go back and three weeks later give in great detail all that was being said to me while I was talking. Another man who must have been rather effective at this operation was Alvin M. Weinberg (1915-2006), now director of the Oak Ridge National Laboratory, and one of the bright young physicists who came on the Project about the same time I did. We'd been close friends and we also had a lot of common interest in the Federation of American Scientists [1945-]. Mrs. Nordheim named Weinberg "Grapeweinberg" because of his effectiveness in bringing physics information back and forth between Chicago and Oak Ridge. We were essentially in a security-hungry area, and we had too little information to do our job effectively except with the personal visits of the leaders back and forth to the Project council.

I think that the security walls impeded the hiring of people except in the field of physics, where the walls were very, very porous. I think, worse than this, that they impeded the forming of political and military decision in Washington about the use of the bomb. (I've talked about this several times in the past.) I surprised Mrs. Smith, in our discussions about this matter, by stating that I considered the Korean War a security consequence of the Manhattan Project.

The whole last phase of World War II had very important decisions that were made by men who were not informed at all about the possibility of an atomic bomb, indeed the imminent probability of an atomic bomb, and so they made these decisions in a technical vacuum. Karl Taylor Compton (1887-1954) told me in October of 19[4]5, when I first met him at MIT, [CDC moved to MIT, in July 1946], that he had called on General [Douglas] MacArthur (1880-1964) on one of the islands in the Pacific in 1944 and tried to tell him a little bit about the bomb, and MacArthur wasn't the slightest bit interested in this at all. Franck said to me in 1945 that he had gone to Washington to try to solicit the response of Conant—I think I told this on the Franck side—about some plans for using atomic energy. He never got such a swift kick so fast in his life, the way he was bum's rushed out of Conant's office. Franck also said that when [Edward Reilly] Stettinius, [Jr.] (1900-1949) was made Secretary of State in 1945, just a few weeks before the Yalta Conference [February 11-14, 1945, Crimea], President Roosevelt asked General Groves to brief Stettinius on the atomic energy project, and Franck says that Groves gave him 20 minutes. He probably got more information later from underlings, but the general story about Groves is that Groves had a total mistrust of most politicians and a great hatred for the British.

He did everything possible to block exchange of information between the United States and the Anglo-Canadian project. He frustrated them continually by telling them they didn't have to do certain things, they could get them in the United States—about radioactive materials, finishing the cyclotron, things like that. They were very bitter about Groves, about the way he treated them individually and his contempt for this project when he went there, and his terrible concern about all these lousy frogs on this project—namely the French. Right at the end of the war, Groves forced on the British government a decision to be put before each Frenchman. The Frenchmen had three choices at the end of 1945: they could leave the project, they could become British citizens, or they could become United States citizens. But he would not tolerate Anglo-French collaboration as long as it was obvious there was going to be Anglo-Canadian high-level exchange of information with the Americans. The French people are very bitter about this.

Lew Kowarski (1907-1979) is a man I know very well who told me a lot of stories about the Groves attitude in the founding of the French project after the war. Of course, the

216

United States government was bitterly opposed to the fact that Joliot, a Communist, was one of the few Frenchmen who stayed in France during the war who had the competence, and he had the love of the French people, and he had a well-deserved Nobel Prize for artificial radioactivity. He had the only cyclotron in France, and he and his students had the most enterprise in chemistry and could do the most work. Madame Joliot had an independent ménage, and as far as I know, there was no collaboration, scientifically, between Madame Joliot and her husband in the late forties. There appeared to be somewhat of a jealousy. She often said in her lectures, "I understand Professor Joliot is making the following discoveries," and she standoffishly presented a poor impression of the work he was doing. She loved him, nevertheless, but I think she was an intellectual rival of his. He was far brighter than she, but very lazy, and she was like her mother, very hard-working but without any sense of humor.

Kowarski was in a position of very high confidence in the Canadian project. He is a Russian refugee who came to Belgium in the early thirties and took his PhD, I think at the University of Ghent, but I'm not sure. [D. Sc 1935, University of Paris, with Joliot.] At least, Ghent was French-speaking. He took his PhD in a French-speaking Belgian university in physical chemistry, and then went to the Institut de Radium, and was a collaborator with Joliot and [Hans] von Halban (1908-1964). Halban is presently a professor at Oxford, I think. [With Francis Perrin (1901-1992)], these three people made the first measurements of the number of neutrons released for fission [1939], and two of these three people escorted out of France in an open boat all the heavy water the French had gotten [in 1937] from Norway, after the fall of France. Kowarski lay on the deck of the boat, on the hatch—the heavy water was in oil cans—reading comic books, while the Germans strafed the whole bunch of boats going out of there. He didn't want to appear excited, lest the Germans think this boat was extra worth shooting at. Then Kowarski worked in the Cambridge project and was one of those being sent to Canada.

At the end of the war, when this Groves' requirement came that the French leave, Sir John Cockcroft is reported to have told Kowarski to put his piles in order and gave him a month to do it, which was a challenge to him to memorize all he could not carry with him. Kowarski has, as I know by testing it, one of the most fantastic memories of any man I've ever seen. He was the key man in getting the French project going in Fort de Châtillon. He's since been squeezed out of the French project, as are most of the old-timers, by the new scientists and the aristocratic scientists from *les grandes écoles*, and he's at CERN [*Centre d'Énergie Recherche Nucléaire*, European Organization for Nuclear Research, on the Swiss-French border], where he says he's second and a half in command, though he's one-quarter time on loan to OEEC [Organization for European Economic Cooperation, founded April 1948 to administer the Marshall Plan], where he has nuclear projects.

He had visa problems coming to the United States in the late days of the old regime, but he's since got things straightened out. He was denied admission to the United States for several years at a time he hoped to get a professorship in the United States. He felt he shouldn't leave Joliot, he told me, while Joliot was in trouble because Joliot had done so much for France and for him, but as soon as Joliot got straightened out or pushed out he felt free to leave the project as it was self-sustaining. His intense application to atomic energy is very understanding, both as to the industrial side and the nuclear physics and chemistry side, and his great wit and understanding of human beings has never been properly realized in the United States.

I'm saying a few more words about the Canadian project. The head of the Canadian project in Canada was the very eminent German refugee [Austrian] chemist, [Friedrich Adolph] Fritz Paneth (1887-1958), who was squeezed out of the University of Koenigsberg in 1933 [by Hitler's *Machtergreifung*, seizure of power, January 30, 1933] while he was on a trip [lecture tour] to England. He went back home and got his papers and left. He was professor at the University of Durham [from 1939] and founded the laboratory of radiochemistry there. When he reached the retirement age for English universities, he was in the same position as many other refugee professors were—he hadn't been in England long enough to have accumulated a very effective retirement fund. He took an appointment at Mainz, where he was made co-director of the Max Planck *Institut für Chemie*. He died in September [17th], 1958, on a visit to Vienna, where his wife lived.

Paneth was a very ineffective leader of the chemistry division in Canada. He was an old, courteous gentleman, very able in the fields that were classic or the fields that interested him, but he never could remember which was neptunium and which was plutonium. And he always had to have his nap, whether a crisis came or not, and, if the Canadian project was one-tenth as hectic as the United States project, there were three or four major crises a day and no one ever took naps or ever took meals at any specified time. There was fight, fight, fight all the time.

He finally left it to go back to England, and the young chemists, which were a mixed group of young British, young Canadians, and a few French, elected Jules Guéron (1907-1990) as chairman of the chemistry division, to their great happiness.

There was no exchange of information known to me between the Montreal project and the US projects from about 1943 on. I think I mentioned Bertrand Goldschmidt could be borrowed by Seaborg but not hired by Seaborg, so he went to the Canadian project. But early in 1946 I gave a talk in Atlantic City, and gave a fair amount of freshly declassified information on the properties of fission products, and about ten young Canadian chemists were down there at the meeting of the American Chemistry Society. It was a program run by Seaborg and he gave a major talk where he mentioned, for the first time, the possibility of denaturation, and there were some other talks—I think I could find the records.

The Canadians were just tickled to death to have information, and they said they would see if they could invite me to Canada. In June of 1946 I was authorized to go to Canada, and I was authorized to take with me the whole body of essentially declassified information on the fission products, which they had seen only on my slides, which they photographed. There was much more information than there was on the slides—there were some 30 pages of data in all. There were about 200 fission products, with about 20 pieces of new information, a lot of which was new to Canada.

I found at that time that two bright young men, William E. Grummitt [Canadian] and Geoffrey Wilkinson (1921-1996), had done, almost without any other aid, a pretty good survey of the fission products, had gotten a pretty good picture independent of us. I encouraged them to publish their material in *Nature* [1946, vol. 163, p. 159] before our material would come out in American journals, since we had several hundred man-years of work against their two-or-three man-years of work. I also was beloved of the Canadians because I had brought a large body of information. It appears that Seaborg and Perlman had been invited to the Canadian project a month earlier, but since the elements 93 and 94 were so packed up with red tape and the like, Seaborg brought no new information. Also, when I visited at the laboratory, I had with me the American project representative at that time, George L. Weil (1910-1995). Seaborg and Perlman also had the American representative, probably the same man, at that time, and they felt [they could not] talk. The most Seaborg would say, if some Canadian would ask if this experiment were worth doing, has it already been done, "Well, I wouldn't work too hard on this field," and he could just hint in what direction to go. But I was in an area less tied by classification, I was eager not to be hindered by classification since I respected the work of these people, so I told them everything I could, although once in a while Weil would say, "Ah-ah, you can't talk about this, this involves the number of neutrons for fission," or something of that type.

I had known Weil from the old days at Chicago [Weil was a Chicago Pile 1 pioneer witness, December 2, 1942] and I don't think I strained the situation. But it was interesting I was watched. Also, while I was in Canada, one of my boys had left Oak Ridge for Boston already and was a graduate student at MIT, and wanted to check with me what he should register for in the summer semester. He phoned to get me at Oak Ridge and found from Grace Mary I was in Canada. Grace Mary had wanted to know how she could get hold of me, and I had told her she could send a letter to a certain official. One had to conceal the fact the United States had a representative in Canada, and one had to conceal the fact that there was major scientific traffic in Canada. Indeed, Grace Mary didn't know enough, and she had to phone the security officer at Oak Ridge who had to wait for somebody to open the safe to find out how to make this letter more safe [secure].

My wife wrote me with the information about [Lionel S.] Goldring, so I could answer him by mail or by telephone, and she had to send the letter to an Army officer but she had to call him "mister." She sent the letter on blue paper, with a woman's handwriting on it. It turned out that the Colonel So-and-so was in Seattle, and his secretary saw this letter from a woman, postmarked Knoxville—there was no Oak Ridge post then, and she didn't think of Oak Ridge—and she thought it was a personal letter, shipped it off to Seattle, and I never did get the letter. But I was annoyed that one should try to conceal so much of the facts in June of 1946 that there was a collaborative effort between the United States and Canada.

I had the pleasure of seeing the Canadian reactor just before it was due to start and crawl all through it, and I saw all of the chemistry the Canadians were doing. Some of it was very, very good. I also found them very eager to hear what I could tell them about what the scientists in the United States were doing about the political-social aspects of atomic energy. Most of these people had feelings very much like those people I knew in Oak Ridge and Chicago, and I'd already come to know, to some extent, in Boston, but there were Canadian laws which involved civil servants not being allowed to participate in public in politics, and they were fool enough to think this was partisan politics, so few of them would make any outside statements in letters to the papers and the like. After my session on the fission process, I was asked by Grummitt to give them about a half-hour discussion about how the United States scientists felt about atomic energy because they were anxious to know what their cousins in the United States did. So I gave them a thumbnail sketch of the formation of the Federation of Atomic Scientists, which at that time was in the throes of being changed over into the Federation of American Scientists. Then there was lots of discussion with the people at lunch and in the evening, though to my knowledge nothing very great came of the Canadian people in public political action although already many of these scientists were involved in visits to the universities and visits to men's clubs and all, giving informational speeches.

Canada is administered very cleanly in science, through the National Research Council of Canada. The leading people in the Canadian project had very easily the ear of those men who directed atomic energy research, and they easily had the ear of the Minister of Development (whose name slips my mind now—he's an eminent MIT alumnus) [B.S 1907] [Don Wiles identifies C. D. Howe (1886-1960), called "Minister of Everything" a dynamo who developed nuclear science in Canada. Email 8 June 2010.] The powers-that-be in the Canadian government sought out the scientists for ideas, how the government should do in atomic energy, and I think the result is that Canada has made a substantial contribution. For a country of about 15,000,000 people its contribution to atomic energy is equally great with the United States. It doesn't involve any military drive for atomic bombs since the end of the war. So there's far less need for Canadians to cry out. At least they took education as a serious obligation; they took the edification and education of government officials for long-term policy.

Session Six — 15 August 1960

Coryell: This is August 15th, and having picked up the tail end of the discussion yesterday, I see that I didn't quite finish my thought.

In Canada, the people associated with the so-called Anglo-French Canadian project which was at Montreal in the Catholic University, but which moved to Chalk River with the building of the lovely little town of Deep River and the building of the fine industrial plant at Chalk River, 15 miles away—the Canadians worried about lectures in the churches, in the high schools and the colleges, and did a lot of the campaign to inform the Canadian public about the value and importance of atomic energy—just as we did in the Federation of Atomic Scientists, which became the Federation of the American Scientists. They also concerned themselves with the dangers of atomic warfare, and the effects of the change of sovereignty.

The Canadian educational campaign is reflected, in some respects, importantly in the far-sightedness of the Minister of Development (his name slips my mind right now) [C. D. Howe] and in the excellent activities in the UN (United

Nations) of Lester [Bowles] Pearson (1897-1972). In the councils of the West, November 15, 1945, Prime Minister [William Lyon] Mackenzie King (1874-1950) and Prime Minister Clement Attlee (1893-1967) and President Truman met in Washington, and around Christmas in 1945 there was a meeting of the foreign ministers of the United States, Great Britain and Canada in Moscow. At that time the Secretary of State in the United States was James [F.] Byrnes (1879-1972). In addition, the Canadians—as I said—had direct contact with the Minister of Defense [Douglas Abbot, August 1945-December 1946, followed by Brooke Claxton, December 1946-January 1954]; they had direct contact with the Minister of Development [probably C. D. Howe], who was responsible for the Canadian governmental efforts in the field. The Canadian government, shortly after this time [1944], made a Crown corporation out of the Eldorado radium and uranium refining corporation [Eldorado Mining and Refining Limited] and took over natural radioactivity and uranium production, and the Canadian government made treaties with the United States and Great Britain furnishing uranium. In all cases, these government officials who felt they were not allowed to enter political activity, that it was partisan, nevertheless carried to the people the educational function and carried to the ministers the judgment and educational function that made it rather unnecessary in Canada for a big campaign.

On the American scene things were more chaotic. The Federation of American Scientists had a big battle that I'll come to somewhat later in these discussions to sell to the public the McMahon bill and to block the May-Johnson bill. Then they had the problem of educating the so-called McMahon Committee of the Senate, which later became the

Congressional Committee of Atomic Energy. But no matter how well-educated this committee is, you have to educate the rest of Congress, you have to educate peace groups, veterans' groups, labor unions, college professors—and there seems to be less common sense and more disorder in American politics, so the educational program in the United States is probably more massive on a per capita basis and has gone less far than it has in Canada.

I remember saying, before my discussion of Canada, a little bit about my opinion that the security problem in the Manhattan Project was related to the Korean War, and I don't remember carrying this argument to a valid conclusion. This is one thing I talked about at some length with Mrs. Smith in our pleasant morning we had together, but I would like to just present the argument again and hear it to completion.

Because of the security barriers, one part of the Manhattan Project knew nothing about the other part, although Los Alamos was, as far as I know, in the position of knowing the technical affairs of all parts. There was a group in the Army and a group at Los Alamos that had control over the production of bombs and knew the production rate. The production rates and the production timing were considered secrets of the highest order, and this is the way I think it should be. It's one area where scientists will really agree in wartime.

The situation was, though, that General Groves was really unwilling to share judgment or share foresight about what the Project was going to do with other people. I think he must have played a more conservative role than Whitaker did with respect to our hot laboratory, in Los Alamos. Every single physicist whom I knew in Chicago in 1942 was sure that we could have a major bomb, given enough U-235. The amount of U-235 was not predictable within a factor of, say, 5. (The Smyth Report says 10, but you take the square root of the answers in the Smyth Report and you get the right answers. The square root of 10 is 3.26.)

With no doubt about the production of the bomb already in 1942, and with the certainty that the factories would produce enough U-235, the atomic bomb was a sure thing and a major thing. It might, however, be too expensive to be a major war instrument in destruction in World War II in case World War II had ended a little sooner or in case the bomb had been held up by some unforeseen technical factors.

On the other hand, Groves never—to my knowledge—let the rest of the War Department, the National Security Council, and the President's staff count on the atomic bomb being used in the war. I mention that Franck said that Stettinius was given minimal information about atomic energy for the Yalta Conference. At Yalta, President Roosevelt worked very hard to guarantee the Russian participation in the East. We paid a high price for this. We offered Russia half of [blank]. We gave them total control of Outer Mongolia, which was then a disputed area. We gave them Germany and Manchuria, [blank], the Manchurian railway. And we offered to let them control half of Korea. Now this is a high price to pay to an ally you don't trust in an area where war hasn't come, where they can get an unassailable position. These bids should not have been made so high on a 90-day schedule [from the defeat of Germany] since it was very likely the atomic bomb would be used in this 90-day schedule. Indeed, the first bomb—the Hiroshima bomb—was used on the 87th day. The second bomb was a few hours before the 90 days.

Russia, on schedule, on noon of August 9th, 1945, 90.0 days after [the] Russian V-E day, entered the war, but by then the collapse of Japan was obvious. [By the time Germany surrendered on May 8, it was May 9, Moscow time.] The atomic bomb probably shortened this, made the collapse come sooner by a few weeks. The atomic bomb was a major factor because it gave a psychological excuse to the Japanese. Well, had the Russians not met the deadlines we bid for, had we not paid the price we paid, Russia would probably have entered the war in the Orient in its own interests if the war had gone as tough as the American planners feared. The invasion of Japan would have been dreadfully expensive of American life, or the life of any invaders. But if Russia had not had the occupation of the northern half of Korea, there never would have been a Korean War. So I give the thesis that the Korean War was one of the prices we paid for excessive secrecy on the atomic bomb; the price we paid for not having the policy makers of the worst war in history know all the facts and potentialities at the level they were known by the most far-sighted and effective of the scientists and military planners.

JBS: This raises the question which Norman [Foster Ramsey] (1915-2011) brought up in his article on the decision to use the bomb on the psychological impact of the Russian entry into the war and the degree to which this, in fact, contributed to the Japanese surrender.

Coryell: To the best of my knowledge, with the high bias of a plutonium man, the decision was made before the Russian entry. The certainty of the Russian entry wasn't known.

JBS: There is testimony from the Japanese saying that, although the Japanese military were able to cover up, really, the Hiroshima bomb to a great extent with the populace in Japan, the entry of the Russians into the war convinced the military forces in Japan that they would have to surrender. They couldn't possibly hope for a more tender peace—

Coryell: I'm not arguing with historians who may know more about it, but I would put the thesis forward that there were Japanese feelers for peace sent to the West by Russia which Russia didn't transmit because it wasn't in Russian interests to have a Japanese peace, that the decision to make peace was a minority vote of the Cabinet and had nothing to do with popular will. It had nothing to do with military will as a whole, but a few key officers. Any intelligent army officer would know that, when Japan weakened, Russia would enter the war.

The new element was the atomic bomb, and I think in all— for farsighted war planning—one would have tested the effect of the bomb, if possible, before you test the effect of public knowledge that Russia was in the war on a certain day. The failure to separate these two effects, the timing of which was predicted to be close together—that is, in the spring of 1945 the bomb was scheduled for mid-July. The Alamogordo bomb could have been the Hiroshima bomb except there were technical reasons for having the Alamogordo test. In the plutonium bomb it was deemed essential to make a test, and the test was so successful the

Hiroshima bomb, the U-235 bomb, was never tested. The Hiroshima bomb is totally different, technically, from the plutonium bomb. I gather from my reading newspapers in 1955 that at least the bombs are quite different technically, and the assurance that we could have the U-235 bomb with enough U-235 is proven by the fact that the first time we had enough U-235 it was Hiroshima and it was successful. The Nagasaki bomb was again the plutonium bomb, and at the time of the Nagasaki attack it was just a few hours before the Russian entry into the war—three hours in Japan time. The third bomb was sitting on the desk of Luis [Walter] Alvarez (1911-1988) at Los Alamos. We led from strength. Bad weather delayed the Hiroshima bomb a few days.

Let me say also that the people at Los Alamos, according to the stories I hear, had a pool about what the size of the Alamogordo bomb would be. The evening before the test there was a major seminar where each one of the project leaders, including your father [Kenneth Tompkins Bainbridge (1904-1996), Trinity test director], specified what he thought the most favorable energy output would be, the least favorable, and the modal—taking into account all factors, there were four of five major areas where the [blank] in the bomb depended. Everybody took a pessimistic view at this last seminar so they wouldn't be caught foolishly. About this time people were making a pool about what the size of the bomb would be. I've been told that only Fermi over-estimated the size of the bomb. Fermi bid highest and he came slightly short of the energy of the bomb.

JBS: No, [Isidor Isaac] Rabi (1898-1988) bid highest.

Coryell: Was Rabi above him? There's also a famous story that Fermi was also impatiently waiting in the dugout with a little wad of tissue paper in his hand he was bouncing. When the shock wave came he observed the displacement and was the first man to compute the energy of the bomb. The major estimate of the energy of the bomb probably came from Sugarman, my old colleague, who had gone to Los Alamos about Christmas of 1945 with four or five of my former Chicago boys, and they determined the efficiency of the bomb by analyzing the debris for plutonium and fission products and could tell what fraction of plutonium had been consumed. This is a tedious chemical problem that was one of my contributions to atomic energy in war and peace.

Let's talk about the petition story. When I read Mrs. Smith's statement and the very nice analysis in the *Bulletin* [Smith 1958], I began to realize how difficult it is to tell what really happened as briefly as twelve years ago. I was sure that I first heard the story of the petitions to President Truman in February or March, 1945, so I read her article very, very critically and cudgeled my memory and came to the conclusion that her dating is probably right. Therefore my time scale was greatly expanded in this period just before the bomb blew up. Things were coming fast and thick, and it seemed that many months passed when it was only a few weeks of pressure.

But at some time consistent with her time pattern, which I would say marks it in early June, Lyle Borst brought me a petition written by Szilard that the atomic bomb not be used in war. This angered me. I felt that we had worked hard with an instrument that would shorten the war, that would give paramount strength to the United States, and would give paramount importance to technical judgments in the planning of the future. The atomic bomb might be horrible, it might kill tens of thousands of people (the estimate at the time at which it would be used), but there was a good chance that one or two bombs would end the war and that the dramatic impact of atomic energy would probably force the world to pay more attention to the world government which was the United Nations. Consequently, the radical change in weight in the world would make man—the atomic bomb would force man to save himself from his own evil impulses and carelessness about war.

This is characteristic of my long-term optimism. In the short term, I felt things would be hell. I was aware already, in the middle of 1945, that there were pressures for preventive war against Russia. My own brother [William Harlan Coryell, Jr.] felt this way. He was a major in the Air Force and had been flying the Hump [India to China via Burma]. I phoned him in the summer of 1945 to check on the health of my mother, who was in her last year of life because of leukemia, and my brother mentioned it was about time that we attacked Russia. I spent about ten dollars of my own money arguing with him about the criminality of preventive war and the touchy situation in the world.

I also had no doubt that the Russians would get an atomic bomb in three or four years, and 3.3 years later—in mid-September [23rd], 1949—George [Bogdan] Kistiakowsky (1900-1982) phoned me to tell me that the Russians had an atomic bomb. He'd been called by the *Boston Herald* to make a statement about it. He wanted to know if I knew what various prominent people in atomic energy had predicted for how long it would take the Russians to get an atomic bomb. I reminded him that General Groves had testified before Congress that the Russians could not do precision machine work and that it would take them anywhere from 40 years to forever to get the atomic bomb. Vannevar Bush had a major book, *Modern Arms and Free Men* (1949), I think, in press which made the statement that it would take Russia 10 or 20 years to get an atomic bomb. Vannevar Bush had the good taste not to make a last minute-change in proof to agree with the fact. He, with dignity, went ahead with the false statement. I heard stories attributed to Conant that it would take a long time for the Russians to get the atomic bomb, but I was contemptuous of Conant's position. I know Conant well, I have a high regard for him as a chemist, I've worked in the same field with him on hemoglobin and heme [groups], and I know a lot of his work in organic chemistry, generalized acidity. He was a fine chemist in his day, but he'd been away from the chemistry laboratory for ten years at this time and as far as I know every decision he was involved in, technically, on the atomic energy project was wrong. He wouldn't release men for the plutonium project because he felt it had no chance of influencing anything during the war. He kept a big stockpile of chemists too long in gas warfare when the Manhattan Project was crying for chemists. Every major project on the Manhattan Project faltered and nearly collapsed because of lack of chemical skill and chemical judgment, and I think Conant was in large part responsible for not releasing men from the overstocked legions of the gas warfare and sleeping powder malaria, and the other very useful projects in their day which were part of the National Defense Research Council.

On the other hand, every man who had had his hands on the Project—men like Harold Urey, I remember in particular, predicted three to five years for the Russians. The people I knew at the somewhat lower levels, who talked less often in public—at least, when they talked in speeches, it didn't get published in the *New York Times*—were of the opinion that the Russians could do it just as fast as we could. I think that Herblock [Herbert Lawrence Block, 1909-2001, editorial cartoonist] characterized the biggest secret of the atomic bomb as "Would it work?" And we proved this. We had to work with the plutonium project not knowing whether there was a positive answer. It was possible that the concepts and the nature would be such that the plutonium bomb was never achievable. If the nuclear constants of the unknown element of plutonium had been different by 10 or 15%, plutonium would have been no better for a bomb than thorium is. On the other hand, it was known already in 1942 that there was no such barrier to uranium.

There's been much discussion about to what extent there were pacifists on atomic energy projects. The famous movie, *The Beginning or the End* [Metro-Goldwin-Mayer film, 1947] that atomic scientists helped to influence some but not enough to be proud of it, where there is a scene of several pacifists resigning to Arthur Compton as the first reactor started on this so-called December 1942 date. I don't know of a pacifist resigning from the Project. [Joseph Rotblatt, from Los Alamos, Palevsky 2000.] I know, however, that many men were very concerned morally about the bomb and hoped that the bomb could not be made, but felt that if it could be made we should make it first, not the Germans, and we should make it before the Russians, and we should give it to the world in a way that would make peace—strengthen international government. I think Franck felt this way. He hoped that the constants in nature would be against us, but he figured if they aren't we have to know it and we have to have a constructive program later.

There were meetings, known to me, going on. They were called Thursday night seminars, I think, in Chicago, in 1944. [Eugene I.] Rabinowitch, Franck, and Hyman Goldsmith were prominent in these, but I never attended one of these. I was in Chicago rarely and I picked up by word of mouth, and I think I picked up sometimes dittoed copies of general, vague statements of theirs that were done outside the realm of secrecy. Things were called this gadget, or this—

The Army objected strenuously to this sort of discussion, and Compton saw to it they were always held out of hours, in classified walls, with minimal writing unless the writing was classified secret. This was a carryover from the pressures growing out of the Association of Chicago Scientists. (I couldn't remember the name of this when I described it two times ago. I think it was the Association of Chicago Scientists. I don't think it was the Association of Metallurgical Project Scientists, because they wanted to avoid this classified title.)

This idea of petitioning the President was brought to my attention when Borst brought the Szilard petition. I resolutely argued against it, that we were in this game for keeps, we were in this game to shorten the war, and we were in this game to use our findings for the good of the world— and I identified the good of the United States with the good of the world. If a bomb like this can be made, the world government, the United States government, must openly meet the issues and plan that not only is World War II ended quickly and cleanly but that we win the peace and we help make a world government.

Borst, I believe, revised the [blank] petition, and circulated it and got a number of signatories. I'm not sure that Mrs. Smith records the Borst petition.

JBS: Did you sign the petition the second time?

Coryell: I wouldn't sign the Szilard petition. I was pretty sure I didn't sign the Borst petition. But a group of us, I might have trouble identifying the men absolutely, prepared a petition that there be an open declaration of the nature of the atomic bomb to be given to the world, and particularly to Japan, before its use and begged that, if feasible, there be an open test—as in Tokyo harbor. I should say parenthetically—I didn't anticipate what is now known, the fantastic contamination that would have happened if a bomb had been fired in the water of Tokyo harbor. But a bomb fired in the sight of Tokyo and Yokohama, high enough in the air that there wasn't much fallout problem, would have been, I think, the civilized, constructive way to expend the first bomb—with a threat spread all over Japan that within so many days there would be a bomb a day or a bomb a week. We could have lied about how many bombs we could have produced. [Perhaps the confusion is a sign of the compartmentation, but in 1995, Howard Gest wrote "The July 1945 Szilard Petition on the Atomic Bomb: Memoir by a signer in Oak Ridge." Charles D. Coryell is the first signature.
http://sites.bio.indiana.edu/~gest/hgSzilard.pdf.]

237

It wasn't until a few days later, after the thing was over, that I heard about the Chicago petition, though it may be that the petition I signed—No, I didn't make a vote, you see. The Chicago vote had been described by Eugene Rabinowitch and Franck and Daniels in the *Bulletin of Atomic Scientists* in 1957, and I understand that the Army raised hell about that. They were furious that there be any discussion about the matter. I may also say that the Army saw to it in this picture, *The Beginning or the End*, that there was a very clear propaganda statement that at Potsdam President Truman had told the Russians about the bomb and that the Potsdam Declaration threatening Japan with death and destruction was a clear statement about the atomic bomb.

When the Hiroshima bomb was dropped, a young chemist at Y-12 whom I know (I think his name is Christenson) made a public announcement to one of the local newspapers that Hiroshima would be poisoned for 50 to 100 years by the fission products in the bomb. We heard by rumor that he was immediately taken into custody, grilled ruthlessly—he had a minor heart attack and he was withdrawn from the Project. It was obvious to me that there was heavy pressure lest any of us talk of poison following the bomb. There was a quotation from Oppenheimer himself that there was no radioactivity left behind that bomb, in the wake of the bomb.

This is essentially true. The bomb at Hiroshima was planned to be at a level it wasn't. However, we were already getting information out of Alamogordo. The Alamogordo bomb was at the top of a tower 150 feet above ground. Even the tower was vaporized. There was the green glass which was dreadfully radioactive and lots of us had pieces of this green glass, called trinitite. It's surprising that the Army let it out, because if one analyzes this green glass and takes into account a little bit of high explosive technology, you analyze the green glass for barium-140 or even today for cerium-144 and for plutonium, you can estimate the efficiency of the Alamogordo bomb. The efficiency of these bombs is still, so far as I know, a rather highly guarded secret. In 1945-48 the efficiency of a bomb was an extremely highly guarded secret. I think I could probably make a public guess about efficiencies now and probably get some criticism, but not get arrested. Had I made a public guess in 1946 or 1947, I'm sure I would have been threatened with jail and browbeaten to death and discredited as much as possible.

When Sugarman came to Oak Ridge to give us a classified lecture on the radiochemistry of the bomb, he refused to answer any questions in this field. He refused to answer them privately. It's the only time I quizzed him about Los Alamos stuff—after all, the war was now over—and he refused to give any information about the ratio of light to heat, what fraction of the radiation of the bomb came in light—things like this would let one calculate anything other than the total power of the bomb.

For a footnote to history, most scientists associated with atomic energy have some of this trinitite. It was a lovely green glass. The surprise about the bomb was that the desert was melted over such a range, and that the composition of the sand elements were such that this was a pale apple green of really great beauty. There was also a rain of red beads— this was the iron oxide in the tower—and these are highly prized. I was promised some by my friend Armand Kelly, who was a mutual friend of ours, and Kimpton, but he never produced them. The beads were much hotter.

Myrna Loy (1905-1993) was appearing on a stage in Hollywood with a tiara made of green glass of Alamogordo. Matter of fact, this is called trinitite—it's a man-made mineral. It came to the attention of someone at Los Alamos—I suspect it was Louis H. Hempelmann (1914-1993)—that this was a serious health hazard. There's enough radioactivity coming from it, and his chief security officer had to go get her not to give the glass up but to take it off her body, when it was found it was quite radioactive.

I've never seen data about the amount of radioactivity that was left in the ground at Hiroshima. I made private computations at the time, based on my own idea of the efficiency of the bomb guess, and I was more concerned about the plutonium in the bomb than the fission products. I still think the government is unduly silent about plutonium. There's a very high toxicity to plutonium, and the ratio of radioactivity of plutonium to fission products passes one at some date that's not many weeks or months after the firing of the bomb. If strontium-90 is bad for human beings, I think that plutonium-239 is in the same ballpark, let us say.

There was no frankness ever about the problem. There's no frankness today about the toxicity of the plutonium bomb, and the frankness of the Atomic Energy Commission about strontium-90 only dates from pressure from Ralph Lapp (1917-2004) and people of that type in the *Bulletin of Atomic Scientists*. Libby explained the terrible contamination from the Eniwetok tests in 1954, the day the *Bulletin of Atomic Scientists* was coming off the press. The government has only talked when there has been massive pressure from scientists who were able to get information free of government classification, so they could get opinions on this from scientists not specialists in the [blank] when they weren't classified, and make an unclassified evaluation and I think this has been a very bad aspect of the United States position.

JBS: Was there at Oak Ridge any possible concern with that at Chicago? I mean, was there any group of scientists who held meetings and discussions?

Coryell: This is what probably ought to take most of the rest of this tape, leaving some smaller rebellions I had in mind out of the picture.

As I can recall now, the petition I first saw was Szilard's and this was countered by other petitions, and we planned petitions and we had a [Wigner] petition, related to the petition at Chicago and the vote. Now, Mrs. Smith has the Chicago story, as far as I know, right. Compton was asked by high authority, presumably President Truman, to get a cross-section of the Project. As I heard the story in the fall of 1945, he was asked to poll more than the scientists—poll a variety of cleared personnel, in a sense to get a cross section of the American public by taking advantage of cleared secretaries and the like. I even heard the story that janitors were included, but I've come to think that this was probably an exaggeration of the man carrying the story. Though it wouldn't necessarily be a bad idea.

As I interpret the poll, the majority of the people wanted a clear declaration if not a test. The plurality, actually, were in the category of no use of the bomb, by a small number. Open tests, with invited Japanese and the United States and Japan, and not using it without a clear declaration. There was a minority that wanted the bomb to be used any way that the military people wanted this, though as I recall now, the largest single vote of the five categories Compton offered, was let the Army use it any way to shorten the war. Partly an accident of the way the thing was worded.

President Truman has later said he takes total responsibility for firing the bomb, for using the bomb the way it was, and I suspect that this is the way it has to be. Apparently, with the aid of Arthur Holly Compton he had some idea of the feelings on this. Now in the current issue of *U. S. News and World Report* is a discussion of the feelings in Washington, and I'm proud to see that one Navy man protested the use of the bomb though most people wanted to go ahead fast.

I also know that there are eminent scientists—some of these are in Boston (one of these told me he was involved in this but told me not to mention his name)—who were involved in selecting what cities would be the targets for the atomic bomb, and the target for the atomic bomb was spared for our bombs so there would be a clear test of the atomic bomb. In other words, the military operations people segregated the atomic bomb from the other aspects of the war, but not on the international negotiation with Russia. I wish the same care in separating the tests had been involved, whether the bomb ended the war or whether the Russian entry ended the war.

The Wigner petition is the one I think I signed. I have no copy of it, but it was one of these that requested a clean declaration of the possibilities of the bomb before it was used, and if possible a test. Shortly after this thing was shown to me and I'd—matter of fact, it was shown to me by a group of people which included Spofford G. English, Arthur H. Snell, Alvin M. Weinberg, Milton Burton, Raymond W. Stoughton, probably Kurt A. Kraus, Ernest O. Lawrence, [Warren] C. Johnson, [L. H. Gevantmann, Lawrence E. Glendenin, and Gordon Johnson signed the Oak Ridge petition], Paul somebody-or-other in biology [Paul C. Tompkins?]. Paul [S.] Henshaw was the biologist [not listed as a signer of the Oak Ridge petitions, there were two, 67 signers and 18 signers of modified petition dated 13 July 1945]. This group of people that I've named—and I think I'd add to this Harrison Davies, Henri Levy [neither signed], [and] I think Norman Elliott—represented a body of people who were very much concerned. There'd be more names than this, but not many more. [67 plus 18 signed a modified version, for a total of 85 Oak Ridge signers.]

We wanted to take some action that would be clean, security-wise, that would be civilized, to indicate our concern with this as a major problem and somehow get it to the President of the United States and to the policy makers in Washington. The Wigner petition was a channel many of us took, and somewhere there'd be a record of who signed it, but within hours after the Wigner petition came to me and I signed it, a phone call came from Martin D. Whitaker that a call had come to him from Groves that this petition was a security violation and would have to be labeled "Top Secret," because the petition betrayed the fact it was close to firing a bomb. They said nothing about the fact that bets in the Army building in Oak Ridge showed it a whole lot closer, because the pool had dates involved and the span of the dates was a whole lot smaller than the meaning of this petition. [Cf. Gest 1995, Smith 1965.]

This angered us, that there was Army interference. We had, collectively—this body of people I'm talking about— had decided we would not let any soldier sign the petition. They were subject to military court-martial; we were subject to firing. We had a number of hot meetings. I'm talking about a date now that represents from July 28th to August 15th. We met on company time. We were mad and we were concerned. We labeled ourselves—or I labeled it—a committee of 26, because there were 26 people. We elected an executive committee, which consisted of those of us who had independent academic positions, so in case any of us was fired by the Army he wouldn't be thrown out of work. Younger men, like Spofford English and Isadore Perlman (I'm not sure, but I think Perlman was in on this) had no independent academic base to fall back on. I was one of those who felt we never could operate anonymously; there must be men who would have the courage to say either, "I wrote that," or "I wish I had and I'm glad it was written."

I think technically I was not on the executive committee, and I don't care. I was in it just as noisily as anybody else. My feelings and my memory are a little fuzzy about the timing of this, but if I recall, it was the petitioning that prompted it. But rapidly, of course—the first bomb was August 6th, the second bomb was August 9th—the Smyth Report was issued on the first Saturday after the bomb was dropped. [Sunday, August 12, 1945.] The first bomb was dropped on a Sunday, and announced on a Monday, and the Japanese news of surrender was expected on Friday or Saturday of that same week. Everybody was confident it would come through the Swedish Embassy, but it didn't come till Monday. President Truman announced it 7:30 Oak Ridge time on Monday, it must be the 14th of August [Tuesday was the 14th in 1945.]

Then there was a rush to get some chance to influence public feeling and some chance to influence Congress. This committee of 26 was shooting in all directions, trying to find a constructive body of things to do, to find contact with the Los Alamos people, the people at other plants in Oak Ridge, the Chicago people, and whatever universities might also be involved in atomic energy. There were no big centers—I don't think there was much going on in New York, but I may have underestimated. There was feeling, though, there was lots of feeling in Berkeley.

I was put in charge, early in the game, of academic liaison. There were also problems of telephone calls. We couldn't send telegrams by telephone. One had to go to the telegraph office to send it. It made me angry because it was one way the Army could control the telegram story better. We had trouble making long distance calls to Los Alamos because of technicalities at the Los Alamos end, but I remember one time I collected lots of 25¢-pieces and made a phone call from the only pay phone in the Oak Ridge National Laboratory, since I didn't want this to go through company wires and get involved with phone tapping unnecessarily. The date for this is probably early September.

I got a letter from Sugarman late in August, which said, "Pals, why don't you put your massive energies in the problem of peacetime policy for atomic energy?" And I was glad to know that Sugarman was in the game. I had all my loose energies tied up in this problem. I was willing to take risks. This worried Grace Mary. In detail they worried her a lot, in general she was very happy I was in this.

I think I want to say a few words about how the news of Hiroshima came to Oak Ridge. Oak Ridge was a tightly held government town. Its gates were (section of tape here is mashed) [blank] no public official, no Congressman, and no newspaperman could come into town without specific clearance from the commanding officer.

There were troubles in this thing—for instance, Waldo Cohn's stepfather [father-in-law by prior description] was a newspaperman but he came in as a father of Charmian Cohn, and this probably offended the Army and may have involved the statement later on that Waldo Cohn had a relative who was on the left-wing press. Mr. [Eblin] visited Oak Ridge several times. He came back to Oak Ridge after the United Nations Conference in San Francisco [25 April to 26 June 1945] and told us his hopes and sorrows about that. As the UN was being formed, a lot of us in Oak Ridge were very much concerned that the UN be formed with the knowledge of atomic energy in mind and we listened with great care to the speeches going on in San Francisco in April and May. The only speech that seemed to bear any sense [blank] with the realities [blank] was Anthony Eden's speech. Stettinius said nothing [blank]. Eden spoke darkly of new forces and new terrors of war, which I think was probably atomic energy.

I'm thinking of a very exciting period in the spring of 1945. I'm going to digress and say I was in Chicago. Early in April 1945, there was a widespread rumor that President Roosevelt was coming to Oak Ridge. I think I talked about this in an earlier tape. Ramps were being built. I was in Chicago the weekend of April [12th, Thursday], [blank] the metallurgical laboratory about 3 o'clock when the news came by telephone. We had no radio inside the laboratories—they were forbidden us. [Blank...] dropped everything. There was a sense of loss as personal throughout the whole chemistry division and the whole metallurgical laboratory as if everyone had had a death of an uncle or a father. I mean, there was a tremendous feeling of reverence for Franklin Roosevelt, because he was credited with having founded the atomic energy projects and the man who had the vision to see that the things were carried on right.

It also happened that I saw Sugarman that same weekend. He'd come into town, I think to get his wife [Gloria], and I put them on the train shortly after we'd heard the news of Roosevelt's death, to Los Alamos. He'd gone on ahead, come back to pick her up, to stay the duration of the war.

When we listened to the speeches at San Francisco and the UN, many of us felt that this was a criminal shortcoming of the United States, that there was no knowledge given to the planners of the UN that there was a new order of magnitude in terror of warfare, a new order of magnitude in diminution of sovereignty, a diminution of the isolation of the seas of North America from Europe and Asia.

These were the things we couldn't talk about at home with our wives. I don't know anybody whose wife wasn't a cleared worker who talked at home about the social implications of atomic energy. [Blank] talk about these are important things, and the job is so important, we've got to shorten the war, and we've got to worry about the UN to save us from all the horrors of world war [blank...] I talked like that to Grace Mary, but she'll remember better than I to what extent I [blank...] I'm still bitter that she knew more than I knew she knew and I couldn't match this.

There was a great deal of propaganda before D-Day [6 June]—I'm going back to 1944 now—about participation— you may [blank...] It was planned that when D-Day was to occur—the exact day was a big secret—that all the churches would be open for all the people to pray [blank...] I took D-Day very seriously, because I always visualized to myself my job compared with the young men strung on the beaches, and that every delay, every insanity of these security people, every idiocy of myself and my own people in not anticipating a problem, I tried to evaluate always, consciously, in terms of how many lives. I knew the death rate per week to a fair degree of refinement, from the official death losses in battles that came in the papers.

When V-E Day [8 May 1945] came, the Army played it down terribly. There was no celebration in Oak Ridge, there was no time off from work [blank...] work harder. They wanted no relaxation because we were close to the climax. V-E Day was close to the critical day when enough plutonium and [fissionable uranium-235] was being made to make Hiroshima, and close to enough [plutonium]-239 at Hanford to make Alamogordo and Nagasaki.

The laundry, which was Army-operated, was an interesting laundry because they specialized in the use of crippled people. Every shirt that came out of there had a big band around it, with big red letters that said "Beat Japan." It was a family joke, that every time you got your laundry you got lousy shirts—not your own half the time—and generally

248

pretty beat up, but "Beat Japan" was always there in big red letters.

I felt so sorry that the sense of accomplishment of V-E Day wasn't given to the people, not for time off from work or for liquor, but for some sense of participation in the war. I also have long since felt that had we had an atomic bomb in time to use in Germany, it would have made a much larger effect in the war than I expected it would have in Japan. I felt a kinship with the German people, I knew that the scientific level in Germany was high, and that Hitler himself had used the propaganda of atomic energy, secret weapons and super weapons. I have the feeling from discussions with Army officials that, had the bomb been ready in January or February of March of 1945, it would have been used first on Mark Brandenburg [Prussian Province] in open heath as a demonstration. We'd have been much less cool about wiping out the lives of Europeans than we were about wiping out the lives of Asiatics.

There was a wide propaganda effect that Japanese were sub-human. They were infinitely cruel and any cruelty returned on them was all right, and that in Asia life counted for nothing and in Europe it did, and that the Germans were salvageable in spite of what they'd done to the Jews and the Japanese were not salvageable because of what they'd done to the people in Nanking [December 13, 1937] and because of what they did to people in Manila [January 2, 1942] [and Singapore]. The horrors and cruelties of Germany and Japan I'm fully acquainted with, now and then, but I always had the hope that there would be occasion to have a public demonstration of the bomb and we missed V-E Day, you see, by just two months.

Late in July 1945, Frank Spedding, my former boss from Chicago and Iowa State University, made a visit to Oak Ridge. I brought him to the house for drinks. I maintained a liquor supply by traveling. I always brought home a suitcase-full, and I brought liquor home for anyone who wanted it if he would stand the loss in case the Army discovered it. I never lost a suitcase. We always had a little bit of liquor to entertain with, because it was very much missed. We were in a dry county, with dry counties around us, and the military reservation was dry. There was no officers club at Oak Ridge like there was at Los Alamos.

JBS: You mean being government property they can't have liquor?

Coryell: Government property can't have liquor, but an officers club can have liquor. There was no officers club. Most reservations do, but for some reason—because of the dry county, perhaps—and both [Roane] and Anderson County were dry. Oak Ridge straddled two counties. The town was in Anderson County. Julie is an Anderson County baby. And the plants were in [Roane] County, as far as I remember.

Spedding was extremely relaxed, happier than I'd seen him for a long time. He told us a couple of jokes. He also made some remark about what he'd seen in the *Denver Post*, about the fireworks explosion in the New Mexican desert. Grace Mary knew what he was talking about, and I paid no attention. This was the Alamogordo test. She never talked with me about it, dammit.

On August 6, 1945, [Sunday, August 5] we had one of the most peaceful days we ever had in the war. There were a number of GIs working with me by that time, probably 25 or 30. One of these, Robert Lesch had against the advice of the Army brought his wife down. Wives were not provided for in the town, so he and she had rented a place in a motel at Cove Lake, about 35 miles north of Oak Ridge. He had to commute 75 miles a day, but I must say that there were a fair number of people who commuted 150 miles a day to get to Oak Ridge. People came from Chattanooga and Dayton, Tennessee. The Lesches had invited us up, and we went up in our old jalopy, a 1936 Chevy. Bob rented a boat and Julie had her first boat ride. We dabbled in the water and were totally at peace with the world.

Some time during this pleasant afternoon and evening the bomb dropped on Hiroshima, and some time before we got back to Oak Ridge, President Truman made his decision to make an announcement on the cruiser [USS] *Augusta*, which came over the air the next day.

The morning of August 7th [Tuesday] meant nothing to me— it was another working day. The sun rises at Oak Ridge at that time of year at 5 o'clock or 5:30, on the double daylight we were on. By the time the 7 o'clock bus got to the plant at a quarter to eight it was sizzling hot, though I think that this day it was a little foggy. It was no lovely place to work—an ugly gray factory with lots of cruddy buildings, not a bit of lawn. The DuPont Company spreads gravel around to save having to cure weeds or plant grass. I call it the main drag, a dirt road, *alamagosso*. [There is no such Spanish word and the only thing close is *alamogordo*, referring to a large cottonwood or Alamogordo, the site of the Trinity test, an

251

area of very muddy *adobe* roads. Joan B. Safford, email, 3 July 2010.]

There wasn't a single paved road in Oak Ridge. All the roads were dirt on which they've poured tons and tons of gravel. As a matter of fact, they were well maintained but the gravel sinks down into the mud. When the mud's wet it's like glue, when the mud's dry it's like cement. It was hell to have a garden at all—we never got very far with that. Some day archeologists will dig through Oak Ridge, and besides finding cement blocks like the hot labs we built, they'll find walls of gravel 10- or 15-feet thick, which is the gravel fallen in the mud. It will be like the eskers glaciers leave behind, and they'll have trouble explaining it.

About 9:30, Ed Brady got news from his wife [Evelyn] that an atomic bomb had been dropped on Hiroshima, and from then on pandemonium broke loose. Everybody was excited. Jane [Evelyn] Brady had the radio going, and had all but the last number of the telephone to Ed's office dialed and every time a news broadcast about the atomic bomb came, she'd dial the last number and either Ed was at the phone or my secretary, June Babbitt, and we got all the news broadcasts as fast as possible.

It also happens that about 9 o'clock that morning a notice had come out that all project leaders were to come to Whitaker's office at 11. I now realize that this was the official broadcast of the timing of the bomb, and he planned to give us the official news as soon as possible. But the news headlines on the East Coast, with an hour advantage on us, were coming in. One of the wives, Doris Edwards, heard the broadcast and walked across the street to the store at Elm

Valley, and met the wife of the security officer who lived next door. She said, "Have you heard they dropped the atomic bomb on Japan?" And the wife of the security officer said, "Doris Edwards, you keep your mouth shut and go right back home and don't talk to anybody."

Within an hour, Los Alamos was mentioned. The news broadcasts talked about Chicago, they talked about Oak Ridge, they talked about Hanford, they talked about the terribly secret place at Los Alamos, they began naming names of prominent people like Urey and all. There was just pandemonium. One of my boys, Don Schover, my electronics boy, had had a radio he brought in to repair, of his own. He had some tough job to repair it, but within minutes he had that thing repaired and we had a radio set inside the plant in violation of security rules and we got our own news without waiting for Helen [Evelyn] Brady's efficient service.

We went at 11 o'clock, the big shots, the chiefs, over to hear Whitaker, and there was trouble with the Signal Corps' fancy radio they'd set up for us, and we got nothing but snarls and jerks, but everybody knew what was going on. Whitaker made a serious speech and he said he hoped that this would be the last time in the history of the world a bomb would be fired, which in a sense represented the hope we felt that this would end the war, though we had yet no news of Japanese reaction to the bomb.

That evening nobody stayed to work late. We went home on the 5:15, the first bus. Harrison Davies, the minute he stepped out of the bus, walked up his driveway hollering

"Atomic bombs! Atomic energy! Atomic power! I can talk! I can talk!"

Tongues wagged in Oak Ridge. We didn't have the Los Alamos parties. Oak Ridge didn't have the delirium that Los Alamos had after Alamogordo, which lasted for a week, but we had about three days of delirium. The *Knoxville Journal* and *Knoxville News* were selling for a dollar a copy in Oak Ridge. They had several pages of spread on the atomic bomb. It happened that at this time Hiram [Warren] Johnson (1866-6 August 1945), a senator from California, died, and he got about two inches. He didn't deserve much more.

It also happened by planned release, the anticipation fuse was disclosed about this time. The war was coming to an end and we could give Army propaganda stories. The timing of ours was set by the timing of Hiroshima, and they never got any news out of the deal. Radar was also talked about a little bit at this time.

That night there had been scheduled a meeting of the chemistry group leaders at Warren Johnson's house. His wife was in Chicago, and Warren phoned all of us—by that time we had phones in our houses—for everybody to bring his wife, and we had a big party and everybody was just thrilled to death. My mother and father phoned me, and the operator was smart enough to track us to Warren Johnson's house about 10:30, and my mother thanked me personally for winning the war. She was thrilled, she now understood why I was so odd in answering the letter I told you about. She visualized me as being the whole story, and I told everybody and there was lots of laughter.

I was thrilled. I was also thrilled by the frankness of the American release, and I was also thrilled by the delicate words President Truman chose. He said, "We have tapped the basic energy source of the universe." I've used this often in speeches—I'm not sure I've paraphrased it right in this thing right now—but when I write it down, as I did in a recent speech on the joys and sorrows of the atomic age, I like to cite the exact wording of President Truman's statement on the [USS] *Augusta*. I felt that the bomb was a tremendous technical advance. It was tragic it was used without warning on the Japanese but, looking at it in a hard-boiled fashion it did shorten the war substantially, it saved Japanese lives. Though I think the United States' position would be far stronger in the world if the bomb had been clearly described in magnitude and horror, and about three or four days were given for Japanese communication. I still suspect, however, that the Japanese stubbornness and human beings are such that the bomb would still have been used, but the United States' position would have been substantially cleaner.

I think, however, that the massive frankness about the magnitude of the effort and the fact that a new era was reached in war and there was great promise for peace—this was all done very effectively the first few days. Matter of fact, I should have said that sometime in June, William [Leonard] Laurence (1888-1977), the science writer for the *New York Times,* had visited Oak Ridge and we were told to talk completely and frankly to him without any security barriers, and whatever he wrote would go through proper security [channels]. I showed him around the hot laboratories and talked to him about the possibilities of having radioisotopes in quantity and what I knew about the medical use of atomic energy that had been shown from the Berkeley cyclotron and the cyclotron work at MIT. I knew a little bit about the work that Robley D. Evans (1907-1995), John Irvine and Clark Goodman had been doing there. I also got a fine impression of the inside of Laurence, and not long after I saw the beautiful piece he wrote about the Alamogordo test, which is Appendix 6 to the Smyth Report. There's also a very graceful [description] which [Brigadier] General T. F. Farrell wrote about the Alamogordo [test]. Of course, we were extremely hungry for this news which came to us now, officially, for the first time.

Not everybody was as naïve as I about the glories of atomic energy and the promise of peace in the world it forced. I had the feeling, and I still have the feeling, that atomic energy forces the world to adjust to the situation, though it may be costly and some more Hiroshimas before this happens. Milton Burton, for instance, couldn't stand the party at Warren Johnson's house and went for a long, lonely walk while he wrestled with his conscience about atomic energy. I don't know how many more people did this.

I know that at my house the babysitter for Julie was June Babbitt, my secretary, and I know that all the younger men in my section were celebrating at our house. I had told June that she could have all the liquor we had, and there was a very happy party of about 30 or 40 young men in my group. I don't know what part Julie played in this thing. She was then 1.5 years old.

For the next few days there was very little science done. But there was precious little rapture about atomic energy. There was much more thinking about what we should have done, and what we should do politically. There was a lot of thinking about what should be the future of the Oak Ridge National Laboratory. There had been a planning committee under Richard Chace Tolman (1881-1948), a man I'd known well at Caltech, and there'd been a subcommittee of this at Oak Ridge headed by Charles A[llen] Thomas (1900-1982). He was at this time vice-president of Monsanto Corporation. He'd been prominently involved on the project of purification of plutonium. This was a tremendous crisis, chemically, when it was felt that plutonium had to be free of oxygen, and fluorine, and some other light elements, to parts per million, and no one could analyze for these. This particular problem was solved like almost every Project problem. A crisis is never resolved. A crisis is supplanted by another crisis. The crisis that supplanted this crisis, technically, is that plutonium-239 has an impurity in it, plutonium-240, which makes more neutrons than the oxygen they couldn't analyze for makes, so they had to design the bomb differently, and that was a Los Alamos problem, not an Oak Ridge problem. But it took the heart out of one of the worst chemical crises I could imagine, an analytical chemical crisis. Remember, analytic chemistry is the oldest part of chemistry, and it's still the toughest. I think I mentioned the boron problems earlier, and what headaches they made.

We worked a full six-day week. Saturday was a full day. On Saturday I was delighted to find on my desk, or in the first mail delivery, a paperbound copy of a report [*Atomic Energy for Military Purposes*, 1945] by Henry DeWolf Smyth, which gave a lot of information about the Project. It essentially gave the planning for the Project as seen in 1942. It gave the predictions then for what was necessary for the plutonium bomb in size, what was necessary for the uranium bomb. It said that the Y-12 facility at Oak Ridge would have to produce one kilogram a day of U-235, which means that it has to have, with 100% efficiency, 140 kilograms a day of natural uranium as a [feed], [which is 560 moles or 560x amperes-seconds processed at 100% perfection (original text is unclear)]. And, if I know anything about electrical engineering, the efficiency was one of 5%.

I guessed then that Y-12 used more power to make U-235 than U-235 itself would give as a power plant, and I guessed at the same time that the inefficiencies of the gaseous diffusion project were probably equally great, although obviously they're no longer great. I think I was wrong on the K-25. The Army closed S-50 on the same Saturday—this was the emergency thermal diffusion process the Naval Research Lab had engineered. The Army closed Y-12 by the end of August. They waited formally, I think, for the Japanese armistice signature, which was delayed until September 2nd.

The Smyth Report was wonderful. It had some bad results, also—it was like the apple of Paris. I got mentioned once, Seaborg got mentioned three times. Milton Burton, who came on the Project two weeks after me, was not mentioned because Henry Smyth by luck took a personnel list from the metallurgical project in Chicago dated sometime between May 4th and May 25th [1942], as a typical project. Burton, who was somewhat older and had been in industry a long time, had come to the Project as instructor, whereas I had come as assistant professor, felt for many years depressed academically because he couldn't cite himself in the Smyth Report. I once resolved that at some time or other I would write a chemists' Smyth Report, to state the chemical balance. In a small sense I did this with the fission products in a speech I gave in Atlantic City in April, 1946. In the publication [C. D. Coryell and N. Sugarman, editors, 1951] of our fission products information we cited the individual authors of the report, which in the section Sugarman had and I had, were the boys who did the work in the lab, not the big shots. My name is only on—there are 350 papers in our book [three volumes]—and I'm only author as a scientist on seven of these. I'm editor of the book. Whereas I guided 40% of the work personally, and wrote all the work in the report form and rewrote it, I figured that I had more than my due share of credit and I wanted the little boys who barely had Bachelors and had to make degrees at least get the credit for their contributions.

Seaborg has done a fairly good job in publications at various times by listing not the personnel that Smyth listed, but listing the pattern of organization at three or four typical dates. So everybody who ever worked there gets on one or the other, and you can see that a man like Brady, who started in my section at the bottom of the list in May, 1942 (he came about May 30th) by September of 1945 he was a group leader at the same level as Harrison Davies. He'd shown his merit as, certainly, PhD-calibre research.

JBS: Was there any reaction to Nagasaki, which took place in the middle of this week? [Thursday, August 9th] Did it change the mood at all?

Coryell: To some degree it was depressing. A lot of us felt that Hiroshima was so well destroyed that give the Japanese a little more time and Nagasaki would not be necessary.

But the fact that Russia entered the war at noon on August 9th was proof to many, including me, that the rushing of the bomb was to beat the Russians and that we were dealing from strength. The rumor came through already in mid-August there were no more bombs in the Orient. [Cf. Gordin 2007.] Also there was the tragic destruction of the cruiser USS *Indianapolis* which carried the bomb and we knew that this was tricky business. It was known in late August that Nagasaki was the plutonium bomb, and Hiroshima was U-235. These things filtered through. Alamogordo was plutonium, but I think the fact that Hiroshima was U-235 was not disclosed by President Truman. These were atomic bombs and that both components are possible, is all that the Smyth Report says. The Smyth Report doesn't talk about 1945, and Truman's statements didn't. Colonel [Franklin T.] Matthias (1908-1993), who was high up at Hanford [in charge of plutonium production], gave a speech in early September, in which he said that the Hiroshima bomb was U-235, never before tested, and the Alamogordo bomb and the Nagasaki were plutonium, and he was censured for this. I soon found that I had to save all newspaper clippings on atomic energy because the Smyth Report didn't say enough to cover all the contingencies one might want in talking to chemists at the University of California, Berkeley or UCLA. I'd lecture to UCLA, or Caltech, or MIT, and I avidly collected these statements, so anything I said which wasn't in the Smyth Report was covered by something I read in the newspapers, so I'd have a security protection. I'm sorry Colonel Matthias was censured, but I think it's valuable to the world to know the information, the important fact being that the Hiroshima bomb was so certain it was never tested. Which means that President Roosevelt at Yalta should have known of this certainty, but I'm sure he didn't. Secretary of

State Stettinius should have known it, Secretary of State Byrnes should have known it. I'm fairly sure that Secretary of War [Henry Lewis] Stimson (1867-1950) did know it.

I might say that when one talks about the pre-war-end philosophy of the atomic bomb, Stimson had forwarded the Franck report to the President. Stimson had made a memo which only talks about the "gadget." It talked about the importance of interpreting this in the councils of the government, and the importance of interpreting this for the benefit of world government. This Stimson report was carefully kept "top secret" for about three years after the war. It was not published till later. This was part of the effort of the big bomb boys to keep public conscience out of this, as was the suppression of the information Christenson gave and the somewhat distorted statement that Oppenheimer gave.

It's true that there was probably no serious contamination at Hiroshima, but this doesn't mean that a bomb means no contamination. We know from the Pacific tests that bombs mean massive contamination unless you're very, very careful. [Cf. Christy, Robert F., Interview by Sara Lippincott, (1994), pp. 90-95.]

The Smyth Report was a marvelous thing for American government and for American science, for it gave a pattern of discussion that was adequate, technically, for most policy decisions. It gave a basis for an information program by the scientists for the lawmakers, for the policy makers of the country, for the schools, for [blank] [the common] man and everything else.

However, there was a strong reaction against the Smyth Report. Many conservative scientists felt it gave the Russians too much information. This was brought to the attention of President Truman and the Smyth Report was withdrawn about August 18th, although it was impossible to withdraw it—the Smyth Report had been passed out to about 5,000 newspapers and certainly the Russians had it. I'm willing to bet a nickel, or a quarter, or a dollar, that the Russians translated the Smyth Report instantly as a classified document and spread it around their laboratories, because the Russians have a love of secrecy and the Smyth Report betrayed the power of American science. The confidence with which one could work in 1942, and the courage to put two billion dollars into a measly two little bombs, and the planning of laboratories effective enough—like Oak Ridge, Los Alamos and Hanford—that they could go on producing for a generation (a generation is ten years in science), or they could go on basic research instantly, as we did already in Oak Ridge and Chicago at odd slacks in the war. We used the fact that scientific information necessary for the bomb had peacetime implications we would exploit out-of-hours.

I was told on fairly good authority—I suspect it came from Warren Johnson, who was very friendly with the Army people—there was a black period of time for a few hours when it was the executive order of the United States that no statement on atomic energy could be made that was not personally cleared by President Truman. This is, of course, an idiotic situation. Truman doesn't know science. This was about the Saturday, which must have been August 18th [correct] or 19th, and then it was obvious you couldn't recall the Smyth Report. It was copied, transcribed, it was all over the world. There was a British Smyth Report—the British committee published the Smyth Report with a report of their own.

Then the Army stubbornly went back. We could stand on the Smyth Report, nothing better, but all of us in the laboratory who were talking out felt we could always stand on the Smyth Report, plus what colonels and up were quoted in the newspaper. General Groves was one of the biggest blabbermouths there was. He said the atomic bomb was so big that only a B-36 could carry it. He said that it cost such-and-such an amount of money. He said so many people were concerned in making it. He was a wonderful source of information we could use to extend the Smyth Report.

In this period of August, we were anxious to talk to Congressmen. We were under security restrictions to avoid going beyond the Smyth Report, and also we were under the restriction that the Congressmen couldn't come in. I just got furious at this situation so I finally sent a telegram to [Horace Jeremiah] Jerry Voorhis (1901-1984), who was Congressman from my district in California where my family lived, though not the one where I lived. I knew my mother and father knew him well. He was the man that [Richard M.] Nixon (1913-1994) supplanted in 1948. He was also on the House Un-American Affairs Committee— Roosevelt saw he got put on as an intelligent liberal, to exercise some counterweight observational power, and he got supplanted by Representative Nixon. I suspect I also wired (California) Senator [William Fife] Knowland (1908-1974). I'm not sure who I wired. I wired Voorhis as a Californian, and I think I wired Senator [Kenneth] McKellar (1869-1957) as a Tennessean, and I also wired Johnson [*sic*]

[Estes Kefauver was Tennessee Third District Congressman from 1939-1949], who was Representative from my district, that there was a substantial block of information closed by the Army and we in Oak Ridge had a great deal of knowledge the Congressmen should have in drawing up atomic energy legislation and we hoped we'd have a chance to talk to them.

I made carbon copies of this telegram and slapped it up on the bulletin board in chemistry and physics, to show that this Committee of 26 wasn't moving fast enough—we had to get out and talk to Congressmen. I think this sort of crystallized. A meeting was held shortly after this, and people moved much faster. Right after this I was made academic secretary.

The problem of public discussion was a very hot one. In fact, anything you said had to be written out in advance and cleared by the Army officer, who was Edgar [J.] Murphy at our group. Murphy was a decent guy about all this stuff, but I was always one who decided to jump the gun. I hadn't been home since I went to Chicago in May of 1942, and UCLA wanted me to come back.

I should say, in this timing, that sometime about the 25th of August, Warren Johnson called a meeting about the future of the laboratory. We drew on the report I was telling about when the company came, the committee of Charles Thomas. I also can't remember the name right now of the elderly man from General Electric that was [blank] committee. Some of us wrote reports—I guess I gave verbal testimony. Johnson wanted to talk to the people. I knew there was lots of restiveness. People thought Oak Ridge was a mess. There was no future in it. It had turned out to be a heavily Southern atmosphere with repressive race relations. No Negro skilled worker in the whole area, although every government contract required there be no discrimination in labor. The Fair Employment Practices Commission [FEPC] was on the back of every contract we made, but not on any aspect of the living of it. There were no Negroes in the school at Oak Ridge. Negroes lived in shabby little huts and were brought in on big cattle busses from the outside [Cf. Fermi 1995].

The problems of the school system worried us. The federal government subsidized the schools in Oak Ridge and the schools were of high standard. They had an excellent superintendent of schools, Blankenship, and Oak Ridge had as good schools as any other part of the United States had during the war, but we doubted this would last.

We had concern about Army ownership of the houses. We had concern about Army control of the telegraph system. We had concern about all of these things. And then people just thought the government was inefficient in the laboratories, and there were a lot of young men who wanted to go get degrees, and the few men that had academic jobs wanted to hurry back, and people wanted to go to cities where they could see plays and go to the theater. I talked about this often, but I never had time in Boston to go to plays and the theater very much.

Already, there were people whose universities were pressing them to come back before the end of the war. Norman Elliott had to give up his position at Pomona, because the president of Pomona College wouldn't wait till the end of the war. This happened to some people.

In this meeting Johnson called, he had planned that George Boyd and Henri Levy and Isadore Perlman, and a few of us talk. He'd arranged, I guess, that I talk first. He just said, "We're going to discuss the future," and that he felt the future was good. Warren Johnson was always a bland optimist, easy-going, slow moving, but he also carried conviction. I just got up and said I didn't know what anybody else wanted to do, I didn't know what my own wife wanted to do, but I intended to stay in Oak Ridge one year, that I thought that the lab had a brilliant future, and that some people had to stay and see it through this period and I thought I'd do what I could for a year before I went back to a university. I think this had a very salutary effect on opinion. It seems to me that the man who says something bold and positive early in the game can dominate the show.

I wanted then to go back to California and see my family again and have them see our child. We made arrangements to go. I also was invited, of course, to speak. I had a job offered at Caltech; UCLA wanted me back. I guess already Monsanto had taken the laboratories January 1, 1945, and people were very unhappy about it. The University of Chicago wasn't running it; Monsanto was running it. (I may be wrong on this date, Monsanto may have come in on July, 1945.)

I liked the Monsanto people, I knew Charlie Thomas, and I didn't like the DuPont organization though I liked the DuPont people that I had to work with. Lots of these were leaving. Matter of fact, many of the DuPont men had gone to Hanford.

Well, I wanted to give a talk at Caltech and UCLA, so I wrote out a bunch of trash—I mean I wrote out what made sense to me, but I never read a speech I've given with any— but once in my life, and I didn't this time. I turned this in to the proper channels and I talked to people, and they said it looked all right. I stood on the principle that the nuclear properties of fission products were not really nuclear properties of fissionable material. It is a fact that the fission products of uranium and plutonium differ slightly, and if you want to be really rigorous the fission products relate to the nuclear properties. But the rules were you couldn't talk about any nuclear properties of fissionable material. Technically, you couldn't talk about the half-life of U-238, but we didn't care, we talked about that.

I openly pressed the issue that it is to the good of the United States that we talk about the properties of the fission products as properties of light elements, not heavy, because they are a hazard in warfare—radiological warfare and bomb warfare—but they are of great advantage for medicine and biology and chemistry and physics. Whitaker supported me on this stand. On the other hand, it's very easy for the Army to say no and it's very difficult to say yes, and I left Oak Ridge approximately September 18th with no decision. (The dates are determinable if necessary.) We went to California on probably a Thursday night—it was a 20-some-hour flight—and it was awfully nice to be home.

While I was there I was got out of bed about 7 a.m. by a telegram from MIT. They wanted to make me an associate professor. I was disgusted. I didn't want to go to MIT and I didn't want another job offer. I had enough already.

I went to Caltech and, lo and behold, there was no clearance for my talk. I sent a telegram to Whitaker which said, "If I don't get clearance for this talk I'm giving it anyway." I sent that off in the morning. The talk was to be given at 3 o'clock. At 2:30 a telegram came in, "Your talk is clear as it stands." So I gave three hours' information in one-half hour of time to a packed lecture hall. Everybody, and my friends, and everybody else packed in and I gave the Smyth Report of chemistry, mentioning names as well as facts. I gave the same sort of talk at UCLA.

I also had been warned by Arthur Compton to stay away from newspapermen and found I couldn't. In essence, I had the first cleared speech of technical scope from Oak Ridge. After the UCLA lecture, we were anxious to go back to dinner with E. Lee [and Midge] Kinsey, the man I got in trouble with trying to hire during the war. There was some reporter from the *Daily News* of Los Angeles whom I thought was somewhat drunk, and was just disgustingly pressing for simple questions. "Can you make an atomic bomb out of earth? Blood?" I patiently explained, repeatedly, what it takes to make an atomic bomb. I'd also said, at the end of my speech—I always ended my technical speech and then spent about ten minutes about the political and human side of atomic energy— I'd mentioned that the only protection against atomic energy was dispersal by cities, that you can't stop a bomb in flight and the bombs would get worse instead of better, I said, "Really, Los Angeles shouldn't be permitted to have more than 100,000 people and New York more than about 50,000." So he said, "Why is this?" I said, "Los Angeles has more than twice the area of New York."

Well, I finally, with family beckoning at the door, finally answered every question he had, and I was just annoyed. The next morning I saw in the *Daily News* a most superb editorial, headed "Little Old New York," that had Los Angeles proud it would be bigger than New York because of atomic energy, and the most beautiful treatment of atomic energy, and a very fine news story about my talk. I think that my patience was extremely well rewarded by this guy. He was trying to make me angry to make a story, or trying to get something intelligent new, as he saw it, to say in the thing. So I resolved I can't avoid newspapermen, the rest of us can't, the best thing to do is be friendly and whenever possible have them check back with me. I asked him to read me the story over the phone; he said he would and he didn't, but he'd played well in having a clean news story and a fine editorial.

Session Seven — 16 August 1960

Coryell: On this California trip I got a chance to assess the feelings about atomic energy of many friends of mine at Caltech and UCLA who had not been involved with the atomic energy project. Without exception, there was a tremendous personal concern that scientists should learn about this effectively, and were grateful for the formal body of information I could give in my speeches and in conversation, and they felt that scientists should do something to influence the government.

At this time it was already known that the Army had prepared a bill, introduced, as is conventional, under the sponsorship of the Chairman of the House Military Affairs Committee, Andrew J. May, of Kentucky, and the Senate Military Affairs Committee, Edwin Johnson, of Colorado. The May-Johnson bill was already public. The Army had been thoughtful in laying plans for this bill—we knew on the Project, I guess it was public knowledge by this time—in having expert technical advice. They had called a meeting over a three-day weekend in Rye (I believe Rye, New Hampshire, not Rye, New York), under Gray, a man thought to be a competent administrator in the Department of the Army or whatever it was called—the War Department. Those involved in the committee included Karl Taylor

Compton, I suspect Arthur Holly Compton, Harold C. Urey, Enrico Fermi, and E. O. Lawrence. I may be slightly confused about this committee, but it's a matter of public record. There was also a separate advisory committee on long-term planning of atomic energy that General Groves appointed, which consisted of four men—A. H. Compton, E. O. Lawrence, Enrico Fermi, and J. Robert Oppenheimer. I suspect there was a tremendous overlap of advisory function.

The committee had given the Army considerable advice, and the preamble of the May-Johnson bill was superb. However, lawyer friends of the atomic groups were responding even as my friends from other parts of chemistry, the nuclear chemistry of UCLA and Caltech, and they had the feeling that the preamble of the May-Johnson bill was a wonderful piece of window-dressing, but that the bill gave no guarantee that many of the principles would be followed. For instance, civilian control was talked about grandly, but there was no reason to assume that Major General Leslie R. Groves, as a retired general, might not be the head of the Atomic Energy Commission. There was very serious concern for secrecy, and the whole thing smacked to the atomic scientists and their liberal friends in the League of Women Voters and the Unitarian Church, and the liberal Catholic ones, and the like, as being a military bill. I sensed this already at Caltech and UCLA from the non-project opinion, and the groups I already knew about at the atomic energy projects felt this even more strongly.

I went with my wife and Julie from Los Angeles to—I went to St. Louis and dropped my family off to see relatives in Leavenworth, Kansas, and participated in a two-day

272

planning meeting at the Monsanto Company, where we talked about the technical future of Oak Ridge and there was very little discussion of scientific politics. Then I proceeded to Chicago, where I had the meeting with Franck I mentioned earlier in these discussions, where the feeling was very, very high. Eugene Rabinowitch had, about this time, started a mimeographed newsletter called *The Bulletin of Atomic Scientists*. There was already a relatively formal organization at Los Alamos, which was called the Association of Los Alamos Scientists [ALAS]. Sometime in September or October, there developed the Association of Clinton Laboratory Scientists.

Martin D. Whitaker, the director of the Clinton Laboratories, protested our using the name, so the thing changed itself over to the Association of Oak Ridge Scientists. We were compartmented totally from the other projects in Oak Ridge, of which there still remained the Tennessee Eastman group. The thermal diffusion organization was dissolved very quickly when the plant stopped operating on August 14th, the day of the Japanese surrender, and the personnel was distributed among the other plants. It had never been going long enough to have much cultural integrity.

There was formed, about this time, an Association of Oak Ridge Engineers, which involved people at Y-12, the Tennessee Eastman Laboratories, and either the same name or a very similar name for the K-25, which was operated by the Kellogg Company. Rather soon in the fall of 1945, certainly before the first of November, these three independent organizations linked under the Association of Oak Ridge Engineers and Scientists—AORES was the name we used for it.

There was at Chicago, a very big Association of Atomic Scientists. It didn't represent total membership of the laboratory as nearly as Los Alamos did. As far as I know, every physicist or chemist at Los Alamos (well, I know one man that didn't belong) was a member of the [ALAS]. The group was extremely articulate, extremely well-organized, but it had no independent contact with the university world and the outside world. And the group worshipped J. Robert Oppenheimer. J. Robert Oppenheimer was formally in support of the May-Johnson bill.

In Chicago I got a sense about this whole program, that there was a split about the bill, and that Los Alamos was crazy. Something had gone wrong with Oppenheimer. People who thought Oppenheimer was a great a man as I felt and still do, felt that Oppenheimer was pulling a great mistake, that the May-Johnson bill was terrible and that it was indefensible of him to be a strong proponent of it in public and in private.

I went on to Oak Ridge with my family after about four days in Chicago, loaded with information from people at the University of Chicago Law School. One of these was Edward [Hirsch] Levi (1911-2000), a man I don't remember meeting, but I had met some of the other people who were prominent in the very early days. I won't go into names now because I don't remember them in great detail.

After time to clean my shirt, I headed for my first visit to MIT. I arrived on a Sunday evening, I think October 20th [21st], to look over the place. I was met at the airport by Professor and Mrs. [blank], and had a pleasant evening. I think I shocked them by the intensity of my feelings about science and public affairs. I also had a chance to see my daughter by my first marriage, Patty, and indeed I saw her mother and stepfather [Gordon Bowler], because they were in Boston. Patty attended a lecture I gave on Monday afternoon at MIT.

I spent most of the morning on Monday, I think October 21st [22nd], being introduced to various people in the chemistry department at MIT. I had known [blank] by having heard him lecture once and spoken a few minutes with him. I had known Professor Miles Sherrill from the Caltech days, but I have the impression that he was not in town. I don't remember meeting him at that time, and he doesn't remember my lecture. Outside that, everybody was new. I knew John Shipman in the Metallurgy Department, and through him I had met Jerrold R. Zacharias, who had recently been at Los Alamos and was scheduled to head a new laboratory for nuclear science at MIT. These people, plus [blank] and John [Clarke] Slater (1900-1976) in physics, who was chairman of physics, and Dean George R[ussell] Harrison (1878-1979), wanted to make a powerful nuclear physics program at MIT, since there already had been one and they wanted to expand it in the spirit of the new nuclear physics growing out of atomic energy. MIT was as much concerned about [blank] nuclear machines, more so than they were about reactors, which didn't alarm me a great deal. I figured reactors were tied up with secrecy.

[Blank] had been persuaded that it would be desirable to have a strong chemistry program. There was a young man, John W. Irvine, Jr., who had gotten his PhD in chemistry at MIT in 1939, who had been a chemist at the cyclotron and they were scheduled to send him down to Oak Ridge for six months to broaden his scope. Then they wanted to bring him back to the Chemistry Department. They wanted somebody at the full professor level. They didn't want to keep Eugene Rabinowitch, who had been a research associate in physical chemistry on a solar radiation project. The reason for this has never been clear to me, and I think MIT let a very good man go, but it may have been the conservative element at MIT were alarmed by his political preoccupations. He has been, since the end of the war, a major scientist in the interpretation of photosynthesis—far more effective, say, than James Franck. He has a gift for interpretive science. He also has been a major spokesman as head of the *Bulletin of Atomic Scientists,* and active in the Atomic Scientists of Chicago, which took the credit for the *Bulletin* founded by Rabinowitch, and a strong person in the Federation of American Scientists. Of course, Rabinowitch is now active in the Pugwash series, the series of meetings which began at Pugwash [Nova Scotia, Canada, 1955], involving scientists from the East and West, and the impact of science in public affairs, particularly in war.

I could also feel that the [blank] were not very happy about my political exuberance, and I didn't care because I wasn't interested in the MIT job. It also was apparent they wanted me as a full professor, because they had found that Chicago had offered me a full professorship. I figured I would be myself, I would luxuriate in the contacts I could make in a big scientific center where I knew practically nobody and where it was obvious, in a matter of four or five meetings, that everybody was concerned about atomic energy. I met Arthur Roberts, I met Louis A. Turner, I met Louis Ridenour

(1911-1959). The secretary of Louis Ridenour was [Beka] Dougherty, who is now Mrs. Martin Kamen.

The enthusiasm was high for positive, constructive action on the national level by scientists. I gave my normal speech, the one that I had forced the clearance on for Caltech, in the MIT auditorium room 6-106, and gave about a two-and-a-half hour lecture in about an hour-and-a-quarter. I didn't realize that MIT always stops the lecture at 5:00 p.m. to let the commuters go home, and very few people left.

I then announced that I had some political things to say, that I didn't want to be part of the lecture, that I didn't want to hold anybody overtime, but if the audience would be indulgent I would give them about a fifteen-minute discussion of what the atomic science groups in Oak Ridge, Chicago, and Los Alamos were doing—since I knew enough about Los Alamos from some phone calls and contacts which had been coming. Mimeographed material had been shipped around and Chicago was close to Los Alamos, so I could pick up a lot more there. Almost everybody in the audience stayed, including Karl Taylor Compton, the president of MIT, whom [blank] had introduced me to beforehand.

Then I gave a ringing attack against the May-Johnson bill after giving a little introduction about how the scientists were very much concerned, and this was a time the scientists had to act in public education and in the direct education of Congressmen. Compton waited until my talk was over. [Blank] brought me over. [Blank] was terribly concerned about my attacking a bill that Compton had helped design. In introducing me to Compton again, he joked about this to make it easier. Compton said, "I was awfully interested in hearing what Professor Coryell had to say." I pulled out of my pocket the analysis by Edward [H.] Levi of the weaknesses of the May-Johnson bill, and told him this had come into my hands just about five days earlier in Chicago, and I'd be glad if he gave me his opinion of it. He also asked if I would make an appointment to see him next morning, and [blank] told me that evening and again in the morning, "For heaven's sakes, Charles, you let Compton talk."

Well, I was delighted to have this friendly talk with Compton. I was taken to his office at the scheduled time, 11 o'clock, and Compton first of all spent about 20 minutes telling me about first his contact with [General Douglas] MacArthur in 1944 and MacArthur's apparent total lack of interest in the atomic bomb. Secondly, the trip he'd made to Japan already—it must have been in September—where he'd seen [Yoshio] Nishina (d. 1951) and other big names of Japanese physics (RIKEN). He reported to me on the state of the Japanese cyclotron art, he told me the recommendations that had been made on atomic energy research control in Japan, and of the parallel job in Germany. Then he spent the remaining half hour telling me about the meeting at Rye [NH] and the various backgrounds of the advice they'd given the Army about the atomic energy act. We also talked a little bit about what MIT should do, and I told him I thought very much it should have a reactor.

The last official act of Compton's life, in 1954, was to authorize a reactor for MIT, but there was a nine-year span in between. I may say also that I feel that Karl Compton is in my circle of gods. I had not many contacts with him in the following nine years, but every one was an extremely warm and productive one, where we—. Several times he asked me for information for speeches of a technical nature, and we were involved in Tracerlab work together, but he was awfully warm to me.

JBS: At this meeting that morning, did he respond in any way to your criticisms of the May-Johnson bill? Was he happy with the way the May-Johnson bill had actually turned out?

Coryell: I didn't feel it was my business to quiz him. It wasn't just following [blank's] advice; I have better sense than most people think when it comes to a really important contact. I felt that my only function would be to bring him the best expert opinion I could, and count on his having the good judgment I suspected he had and I later knew he had. He's the best administrator I know that science has produced, and I know personally that he was still very effective thinking in scientific channels, though he failed in his effort to do research after he became president of MIT.

He told me this story himself. He brought a man with him and set up a spectroscopy lab in the basement of building 6, and he planned to get there every Thursday afternoon. He got there twice, and he dropped the myth after another year or so. But he had a keen interest in science and he had a creative mind, and he had appreciation of new trends in science. He was effectively, for my mind, an able scientist, and on top of this he's the best that I know of the scientists produced in administration, and it's produced precious few. Comparable with him are only [James R.] Killian (1904-1988), who's not really a trained scientist, but who's equally effective, and Julius Stratton (1901-1994), and both these men learned tremendously from Compton. I think the strength of MIT comes from these three presidents, plus some of the past presidents like the founder and [Richard Cockburn] McLaurin (1870-1920) who, I think, were giants in science and administration. I would have liked to propagandize Compton, but I would much sooner hear everything he had to say.

I left MIT, I think, Monday morning. I didn't go to MIT, I went to Harvard, to call on an old friend of mine, John T. Edsall (1902-2002), whom I'd known before the war at Caltech, a biochemist, a man who is the picture of dignity and probity, a kindly, fatherly man, much older-looking than he is. It happens I had met him in Chicago a few weeks earlier, and we had participated in sending a telegram protesting some statement of [Fleet Admiral Chester W.] Nimitz (1885-1966). The details now elude me, but Nimitz made some statement about scientists in Japan, or scientists in the world in respect to the atomic bomb, that was not only nonsensical, but irritatingly wrong and dangerous. I picked up at random three or four scientist friends of mine in the street, and we sent a hot telegram. Perhaps before the end of this series the contents will come into my mind, but I knew that Edsall was keenly interested. Edsall introduced me to Louis and Mary [Peters] Fieser (1909-1997), whom I later came to know rather well with respect to Israeli connections, and I found that the temper at Harvard which has had no direct participation—no, they did, through Los Alamos, but I didn't realize it. The chemistry people at Harvard, and the biochemistry people at Harvard, so far as I know, had no participation in atomic energy programs. The physics department I had to learn about later.

I caught an airplane for Washington. I remember getting a *Boston Globe* and seeing an editorial about atomic bombs being like pregnancy, you can't have half of it, and called that afternoon on Leo Szilard, who had moved to Washington to run the underground campaign of the scientists against the May-Johnson bill. I knew about my MIT trip—I planned it shortly after a telegram came to me at Boston, I had to talk by telephone to Arthur [blank] from UCLA, and so I had written to Senator Knowland, as a Californian, and I'd written to Senator McKellar, as a Tennessean. I guess I already mentioned I'd made dates to see these—did I tell this?

JBS: You didn't say that you'd made dates to see them.

Coryell: I said I'd expected to be in town about the 23rd, and if they didn't mind I'd like very much to have a few words with them about atomic energy. Oh—I mentioned the names previously in that I'd sent telegrams about the terrible information situation and how important we thought it was, so probably my name was somewhat known to the office attendants.

I can't remember where I ran into Szilard, except it must have been in the House Office Building. Szilard had the habit of ingraining himself in the oddest places. It wasn't in a hotel I was talking with him, it was in the office of some Congressman. What Congressman I don't remember, possibly [Chester E. "Chet"] Holifield (1903-1995) of California.

Voorhis took me to lunch the next day, and introduced me to Emily Taft Douglas (1899-1994). Both of them were interested in what I had to say, particularly Emily Taft Douglas from the Chicago district, and Voorhis because Caltech was close to being in his district. Later on, Voorhis was to be very, very helpful to Lyle Borst, when he was Oak Ridge representative in Washington about a month, and Borst went to Voorhis because of the warmth Voorhis had shown me.

I can only use the word flabbergasted for the reaction Szilard showed when I told him I had appointments with Senator Knowland and Senator McKellar. In fact, by that time I think I had an hour appointment with McKellar at 3 o'clock the next day, which would—by the present line of reasons—be Friday, the 25th [26th]. No, I'm not sure. It could have been Thursday, the 24th [25th]. I think I was home and collapsed with tiredness by Friday the 25th [26th].

This story is already on the card, is it on the meeting with McKellar?

JBS: No.

Coryell: Szilard said, "How did you get this?" I told him. He said, "You don't meet Congressmen this way. You go to a friend who has a friend who knows someone to take you there." Szilard was making his connections through a labor lawyer I never met, a man I think whose name is [Robert K.] Lamb. It was unheard of to go to McKellar, because you couldn't trust the man.

I said, "Well, what do I do? Cancel my appointment?" He said, "Heavens, no." He said, "I want you to check with me just before you go. Senator McKellar is dean of the Senate.

He has the responsibility of appointing a chairman for the new committee Congress has authorized by the bill from Brien McMahon. Normally the man who proposes a committee is made chairman as a matter of courtesy."

Szilard described to me that the Senate had, just a few days earlier, passed the McMahon Resolution to appoint a Senate committee on atomic energy. By normal Senatorial procedure the man who proposed the committee would be made chairman, but in this case there was massive competition. McMahon was the junior Senator from Connecticut and Senator Edwin [Carl] Johnson (1884-1970) [from Colorado], of the Military Affairs Committee, a man who was said to hate the Army, thought it should be his committee, and Senator Tom [Thomas Terry] Connally (1877-1963), from Texas, who was the head of the Foreign Affairs Committee thought it should be his business because of the importance of atomic energy in foreign affairs.

Szilard told me to check who was the committee chairman. If McMahon was appointed, I was to give effusive praise to McKellar; if Johnson or Connally had been appointed I was to "bitch like hell" (these were his words). In any case I was to give him massive numbers of scientists who were voters because he only paid attention to the size of the vote. If no appointment had been made, I should give every pressure I could give, short of telling an absolute lie, to get Brien McMahon in. I felt that this would be a tough job, but at 2 o'clock no appointment had been made. I went to the Senator's office and was met by a crusty old gent who looked at me over his glasses and said, "I'm Don McKellar, the Senator's brother. The Senator isn't feeling very well this afternoon, and has asked if I would talk with you. He'll see you later, tomorrow, perhaps, if it's important." I said I thought the matter was important, but I was quite satisfied that he would transmit the information to the Senator.

I told him that I came representing a scientific group from Oak Ridge, where I was living, from Caltech, from UCLA, from the University of Chicago, from Northwestern University (because I had met one Northwestern man at Chicago), from MIT, and from Harvard, and that I had reason to think that the scientists in all the universities in the country felt much as I did. It was a matter of great urgency that we have an effective committee in the United States Senate under an able, young man who would have a chance to learn about science, and that we have an atomic energy act that gave proper position to atomic energy and its peacetime developments and didn't make it a tool of the Army. The Army had made enough messes in the war.

He asked for a few more details, and he asked me to give the names and the addresses of the officers of these societies. I listed a friend for each one of these places, and the first thing I had to do on leaving Washington was to cover my ground. The man I listed at MIT I won't name. He's a sweet, lovable man whose heart's in the right place but hasn't the courage to say no to himself and he would have been paralyzed with fright if a letter had come from McKellar asking is it true that 300 scientists at MIT feel roughly what Professor Coryell has reported for them?

I then spent a little time kidding him about Oak Ridge as being a Democratic island in the Republican end of Tennessee and chiding him for not coming to visit the place and offering to show him personally through the town—I

284

had a pretty old jalopy, a 1936 Chevy, but I knew Oak Ridge pretty well and I thought it was a place that was politically very alert and would be an important part of the liberal Democratic movement. I knew very well that McKellar wasn't liberal, but he was interested in anybody that voted Democratic. We left very cordially.

I think I'll finish the McKellar story right now. I want to go on and talk about the evening that was arranged by [Harlow] Shapley (1885-1972) and Watson Davis (1896-1967), which was also tremendously thrilling. I certainly felt wonderful as an American citizen, to step into the Senate and House Office Buildings and be treated as an important person because all I could see on my side was that I came from a long ways off. But I think this is true, that most of the Congressional offices try to make the constituent feel he's important, and while not everybody spends very long with the big shot I'm not at all sorry I didn't see McKellar directly, though I would have liked to have the pleasure of talking with the man before he died.

I felt that my message was his if it was important, and statistically these messages are handled as if they're important. They're weighted against the other forces, and I was just lucky to be in on the day decisions were being made and when scientists were still very rare in Washington.

I went to Oak Ridge the next day, and was ill and exhausted. I guess I got in Friday night. I don't remember how I spent Friday morning, possibly on a little sightseeing since I'd only been in Washington a few hours before in my life. I heard over the radio that evening that Senator McKellar had appointed Brien McMahon chairman of the Senate

Committee on Atomic Energy and I felt obligated to send a telegram of appreciation for this. But I felt so miserable that we were not allowed to telephone telegrams—I would have had to get out and go down to the telegraph office—that I thought, "Oh, I'll take care of it in a letter Monday." The next morning a night letter came from McKellar, which said, "You will see that I have appointed Brien McMahon chairman of the Senate Atomic Energy Committee." I did send McMahon a letter. Apparently the luck of my arrival with information did a lot of good.

I should also have said that either Thursday morning or Friday morning the newspapers carried a story from Boston that there had been a meeting of 300 scientists—if I recall, it was in a new lecture hall of Harvard University—which gave by acclamation a resolution stating that, without prejudice to the May-Johnson bill, they called for open public hearings on atomic energy. There were 300 signers, but they listed only the top five or six. It began with Karl T. Compton and [Leonard] Carmichael (1898-1973), as president of Tufts University. So here again, the Edward Levi move and the timing of my trip were just critical. I have no doubt that Compton would have taken the same position without my pressure, but he might not have been quite so fast in taking it.

JBS: While you were in Washington, did you see Senator Knowland?

Coryell: Yes, I spent about 15 minutes with Senator Knowland—I think right before I saw McKellar. Knowland had only been in the Senate about 10 days. He was an appointee to fill in the position of Hiram [Warren] Johnson

(1917-August 6, 1945), who died about the day the [Hiroshima] bomb was announced. Senator Knowland struck me as a severe, earnest man, and I formed an opinion of a man who was concerned in trying to learn about this sort of problem, an approachable man. I have never had any contact with Knowland since then—I think I wrote a note to each person who saw me, pointing out what had happened since, and thanking them, because I figured that if I weren't in Washington any friend of mine could exploit these things.

I also dictated to my secretary, June Babbitt, a five- or six-page odyssey of Charles Coryell, which outlined what all I'd done of political import, whom I'd seen, whom I'd quoted to whom and all, and this was sent to all the people involved, so that if anybody was caught they'd know what I was like even if they only knew me from a few minutes contact at MIT or at Harvard and would know the web of commitments that scientists were freely making back and forth.

We'd already learned in Oak Ridge that things moved so fast that one had to trust the good judgment of people, and the only thing was you had to report back what went on so that in case you made a mistake someone could cover for you. Very shortly I was offering three or four people the privilege of signing my name to any petition or any document if they also would sign and had checked with at least two other people among this group of 26, figuring that the thing was to exploit rapidly the unity of our position because our position was very much the same in all the groups. There was almost a disgusting monotony about the memos each one of us created and passed to the others about the threat of the atomic bomb and the necessity for world government and the nature of the political control and the nature of the technical control possible.

Some of these came in anonymously from Los Alamos, and Chicago and Oak Ridge, and I had the strong feeling that the Army would chop us down for any anonymous documents. Any document that came by we replicated on our mimeograph machine. We did them out of hours, we used government paper and ink, but not government time. I always attached my name or someone else's to them, not to claim authorship but to claim political responsibility in case of a charge of a security slip, because I already knew from the Wigner and Szilard petitions that anything could be declared top secret. This was a weapon available to any Army officer to embarrass or hurt us. We also never let any soldier bring his name out. Before we went on, we asked their opinions as private people, but we never put their names on because they [were] under a formal discipline that was absolute.

About a month later, Drew Pearson (1897-1969) reported in his column about Milton Burton and Congress. Burton was part in Oak Ridge and part in Chicago at this time, a very dear friend of mine. He reported that Charles Coryell had kidnapped a relative of the Senator and beat him over the head until he promised to make the Senator appoint Brien McMahon. Now this is an exaggeration of the way I treated Don McKellar. I think Don McKellar died sometime that fall or that winter, and [Ken] McKellar was to die not more than a year or so later. [Not reelected, 1952.] I felt seriously embarrassed by this, but I also felt delighted at the opportunity this might give me to meet Drew Pearson. So I immediately sent a letter to Senator McKellar, telling him this was an obvious falsification situation with no bearing to the truth, and that, if I wanted to impress the Senator I wouldn't do it in the newspapers, I'd do it by direct contact by going to him. No response ever came to this note.

I wrote Drew Pearson the story, telling him a little more about what happened, and telling him that this might damage our relations with McKellar but that it made awfully good reading, and I also got no response from that. I also have never met Drew Pearson, though I still read his column with interest. Pearson was inclined to be very friendly towards the atomic scientists; he gave us lots of news.

Raymond Gram Swing (1887-1968) was another person that Harry Brown made a contact with, and was very effective. When I say Harry Brown made a contact, he made a contact from Oak Ridge. Contacts were made all over the country and anybody shared a contact with anyone else.

We also found, in the transmitting of documents back and forth, that there were other groups besides atomic scientist groups. In our mind, Cambridge, Massachusetts, was radar. This was the bulk of the major war activity and there were lots of people involved, but there were, as I now know, lots of people from biology and medicine that weren't this, and there already were returnees from Los Alamos. Major scientific figures with major obligations, like Kistiakowsky or Bainbridge or others, may have come back already in September or October, whereas people like me who weren't indispensible to the departments would linger on to see the job through and people without academic appointments in 1945 were waiting to see which offer made the best sense.

There was a group at Cumberland, Maryland [Allegany Ballistics Laboratory, 1944-], the Association of Allegany Scientists, and this word Allegany is spelled differently from the Allegheny River and Mountains. It's spelled after the town in Maryland [West Virginia] where the rocket group was. I think that one of the people active in forming this was probably Alexander C. Kossiakoff (1914-2005), who was a brother-in-law of my dear friend Harrison Davies who had been one of the bright boys at Caltech when I was working with Noyes, and who was in my group. He worked with [blank] silver as an undergraduate research project, and I helped write up his stuff in the Noyes and Kossiakoff papers that were published in the old days. I've seen Kossiakoff only a few times since. Later on Kossiakoff went into government science and, as far as I know, stayed out of noisy politics. I think his heart's in the right place but he doesn't necessarily, as far as I know, give speeches or write Congressmen. I know he's director or associate director of a laboratory, but I just don't know what his politics are.

There were other people in the rocket field, proximity fuses, and things like that. By the way, I used the word anticipation fuse instead of proximity fuse. Well, as I say, there was a moderate amount of monotony in all this stuff, but it was evidence to me of the unity of the scientists who'd had a few years to think about the new weapons. The rocket people told us there'd be no question that a rocket shot 300 miles would be accurate within a mile, so rockets with atomic warheads would do tremendous damage. This is what people like Kossiakoff told me was the technology in 1945.

The proximity fuse gave some indication of how the Hiroshima bomb was fired at a certain height in the air in Japan. It's obvious that the atomic bomb technology involves a lot of high explosive technology and atomic bomb delivery would be far worse when it had missile delivery, and that these things were major research programs. But it's also interesting that the scientist who worked as far away from the atomic bomb as this was eager to know about the atomic bomb and to warn the world that the weapons of war had become much more serious, that sovereignty and isolation were diminishing greatly, and that the methods of offense had taken a leap of several orders of magnitude with no prospect of any move up in defense. One couldn't talk technically of a stalemate between offense and defense. You could talk only of the deterrent of the threat of overkilling to offset the threat of starting a war.

The other thing is that, almost without exception, the political answers envisaged by these young scientists, few of whom had had, probably, little political science and government (I'd had none either), involved making the UN strong or found a world organization that would be as good as the UN should have been but wasn't, with the atomic bomb as part of the thing, with no veto, with weighted representation according to population, technology and education. There was, in the fall of 1945, planning of quite a number of organizations either competitive with the UN or with the UN so modified it would be a new and much better organization. Robert Maynard Hutchins headed such a group. There was another group that my pacifist friend Don [DeVault] was involved with, which planned on adding only a government that was responsible for people, and they'd count on getting popular votes here and there throughout the world, and assess popular reaction to world government.

I was quite interested in all these things, though mostly I played it fairly conservative. I figured the UN isn't awfully good but ought to be made better; I'll stick with it. But listening to all these other things [blank...] UN, and I'll see that the UN is changed much more slowly. There should have been a charter review in 1956 and there wasn't. But the UN moves by rising to crises only and on the whole it's risen fairly well to crises, sometimes with a little delay—not in the case of Korea, but definitely in the case of the Congo, for instance, or in the case of the Suez crisis in 1956.

At any rate, these gave me firm conviction that the scientist had to be present in politics to help influence and educate government lawmakers and executors, and I also felt intrinsically from my old Caltech days that science should be part of this. I felt bitter, and still do, that there is no honest scientist in the United States Congress. There are a few medical people, there is one chemical engineer whom I, of course, corresponded with several times, but there's no sign to me that he has ever spoken as a technologist or he has acted in technology. He happens to be a man who made a living in chem[ical] engineering. He's representative from Los Angeles, and I think he's still there. Walter [Henry] Judd (1898-1994) is a medical man [from Minnesota]. I met him also, through my next-door neighbor, Lib Taylor [cf. Matt Taylor], sometime in the fall of 1945 and had a very effective conversation with him, but Walter Judd, I think, is much more occupied with China as seen from the eyes of an ex-missionary, with medicine in the genre of good will than by whatever science is involved in medicine. Now, some aspects of medicine are highly scientific and others aren't.

The highest standing scientist in Washington in 1945 was Edward U. Condon (1902-1974), who was the newly appointed head of the National Bureau of Standards (1901-1988) [then renamed National Institute of Standards and Technology, NIST]. I had never met him until this day I saw Knowland and McKellar, but Szilard brought me along to a dinner at one of the fancy hotels, probably the Statler, which seems to have been the second in a group of three or four dinners of scientists and senators. They'd intermingle, have cocktails and dinner together, and have a friendly, off-the-record discussion in the evening of what problem the atomic bomb offered.

To my great delight I was invited to this dinner, and then the headaches began. A tremendously familiar face—a man waddled across the scene with a tremendously familiar face, was introduced to me and greeted me warmly, but it took me a minute to recognize Senator [Arthur H.] Vandenberg (1884-1951) [from Michigan] from the cover of *Time*. A long horse face to my right that was hard-thinking proved to be Senator [Leverett] Saltonstall (1892-1979) [from Massachusetts]. Senator [Charles W.] Tob[ey] (1880-1953) from New Hampshire cornered me and pressed me. First of all, I had to tell all about my family and children and grandchildren, and then he wanted to know what proof scientists offered for the existence of God. I didn't want to lie to Senator Tobey, I didn't want to hurt him, I didn't want to be trapped, and I think I satisfied him.

I sat down across the table from Senator [Irving McNeil] Ives (1896-1962) [New York State Senator from 1946, CDC error compared to his "Odyssey" notes, (Smith 1965, p. 190)], with Senator Saltonstall on my right. There were a few more senators and scientists. I think I was at the corner of the table, so on my left was nobody. The leading scientists there were Watson Davis, of Science Service; Harlow Shapley, who was running everywhere; Szilard, who was running everything but not running everywhere; and J. Robert Oppenheimer, whom I had known a little bit before the war, he was one of the Caltech and Berkeleyans whom I worshipped and still do.

After dinner, Shapley spoke a few minutes to say what the score was. Shapley was master of ceremonies, and it was my first meeting with him, but I have a wonderful regard for him. He's a lousy speaker, but he has the right ideas. He sells them anyway. He's a lousy speaker because he never prepares his talks. Oppenheimer, as usual, made some very pertinent comments in very contorted sentences, but struck deep into the heart and mind together. He has a gift for seeing problems. I've seen transcriptions of his that are awful, and then I've seen, of course, the prose he writes. The transcription may seem awful, but if you hear it or read it, it strikes deep. His speech is contorted because he seeks carefully the words that say what he wants to say, knowing the weight and value of the words and not worrying about the accidental grouping of the sentences and paragraphs.

Senator Saltonstall made the most able impression to me of the senators. I told him that I was considering a position in Massachusetts and would therefore be a constituent of his if I made the move, and he gave me some arguments for it. We talked a lot about the Boston scene, but whenever it came to a situation where it was obvious what the Senate should do, Saltonstall had the answer effectively and it seemed to draw the interest of the senators. He kept protesting, however, he didn't know one isotope from another. His technical knowledge was almost zero, but this of course was true of everybody there.

Across the table from me [Senator] Ives was a total bust. Ives would carry along with the conversation among the senators about what the Senate should do, then all of a sudden he'd say, "Just a minute, boys. This is election year and I've got a tough job." Anything that involved intelligent courage Ives always went along with to a high point, and then dropped, reminding that roughly half these men were facing the election in 1946. Ives was defeated. I'm not surprised. Senator [Glen H.] Taylor (1904-1984) [from Idaho] was there, without his guitar.

Shapley wanted all of the scientists to have something to say, and just after Senator Tobey had made a ringing statement of 10-minutes' oration and peroration—you could hear the American eagle flap his wings and everything else, all for the *Congressional Record*. The senators spoke in the period as though without any change or proofreading it could go right into the *Congressional Record*—I got up and made some remark about the well of goodwill (I quoted Wendell Wilkie, (1892-1944), because I was in an audience which was about half Republican, half Democrat) and all of a sudden I realized I didn't have any peroration, I didn't know how to stop this thing. I immediately sat down and felt like the most miserable man in the room.

295

But I was walking on air after [seeing] this cross section of the leadership of the American Senate. I also, from my contacts that day and the day before, came to the conclusion that there probably are very few Senators or Representatives who are idiots; matter of fact, that the IQ [intelligence quotient] of the United States Congress probably rates as high as the IQ in colleges. Well, there's not much place in the Congress for genius and some colleges will tolerate genius, but many colleges I know make it tough for a genius.

These men operate under tremendously different constraints from college teachers and scientists. They have to think of election, they have to think of the *Congressional Record*, they have to throw their attention to whatever comes up, and they have to worry about party policy and the like. So their actions don't come out as we see it, the actions—collectively—are certainly compromise actions, and the actions of any individuals are. But I never want to underestimate the ability at thinking, and the ability at expression, and the ability in justification for its position that probably almost every member of an important government body of legislation has.

I also have long since found that, if I differ from a senator or representative or member of the Atomic Energy Commission, I should write him with the assumption he knows what he is doing, but he may not know all I know, and I try to treat him as a gentleman and give arguments to change him, not insult him or just make a provocative statement or an "I know you're wrong." The result is that in almost every case I've differed with a Congressman, I have gotten a letter that was pertinent to the situation and represented justification for the man's position which helped us better counter it if we found we had to. In some of the cases, I've been able to feel that a personal letter from me or someone in our group—we often prepared these letters and talked them over among ourselves—has made a substantial change in position.

JBS: Do you remember, specifically, after dinner what the subject of the science talk was?

Coryell: Well, the point was the scientists were there to educate the senators in how important atomic energy was. Shapley gave a little talk about isotopes and power, and this sort of stuff. We were to be at their service for technical know-how, but we were there to show from our conviction, based on four years of experience, how urgent the problem was and find out from them how we could help the Senate achieve a rational atomic energy act and put the country in a rational position for the international control of atomic energy, the national development of atomic energy for peacetime purposes.

Brien McMahon was certainly there, but I didn't know at that time he was going to be made chairman. Sure, he was there, but he didn't stand out in my mind because he was just a Szilard problem. I don't remember whether Congressmen May and Johnson were there, but I won't guarantee they weren't. As I said, this seemed to be the second of such, and there were planned to be more. I don't know more about it, I was an accidental arrival in Washington and extraordinarily lucky to be in on it. I was an accidental arrival in Washington with enough information that McKellar thought it was desirable to tell me that he had reacted to my advice, although there were certainly other pressures on McKellar than my own and I didn't make that choice, I'm sure. But it was convenient for him let me think I did.

I think I probably should tell one more story about the United States Senate, which is, so far as I know, my last story unless something in the Pauling case comes up. Brien McMahon quickly learned a great deal about atomic energy, and I gathered since his death (he died of cancer in the fall of 1952—he hoped to be a vice-presidential candidate in 1952 but he was never able to go to the convention) he was substantially influenced by a former law partner of his, Gordon Dean (1905-August 15, 1958). (By the way, yesterday was the second anniversary of the death of Gordon Dean in an [Northeast Airlines] airplane accident over Nantucket. I came to know Gordon Dean rather well after 1954.) Brien McMahon, through people like Condon, Shapley, and Oppenheimer—this group here—sought out technical advice, tried to inform himself and became rather well informed. He led a liberal wing in the committee for atomic energy.

I don't remember when the committee for atomic energy of the United States Senate became a joint committee. This is a matter of record, but my Washington contacts were limited. I haven't been in the halls of Congress many more times than this one glorious time in October 23rd, 24th, and maybe part of the 25th, in 1945 [week of 22-26]. The Senate Committee for Atomic Energy included some of the people in this room. Vandenberg was on it, I'm pretty sure, and I suspect [Bourke B.] Hickenlooper (1896-1971) [from Iowa] has his seat. It did a lot of self-education and went visiting.

Sometime in the middle of November, 1945, a visit from them was scheduled to Oak Ridge. About pulling wires, Szilard [blank] all others, the Association of Oak Ridge Scientists, tried to get the Atomic Energy Commission to be guests for dinner. General Groves did everything possible to block all contact with this committee, and thank God General Groves was obstructive. I don't know who made plans for this affair. Fundamentally, nobody at X-10 was very closely associated with anybody at Y-12 or K-25, because the technical compartmentation had brought with it a social compartmentation. So no one could speak for Oak Ridge as a whole. I think Warren Johnson may have had a lot to do with this. Warren Johnson was director of the chemistry department but he played poker and bridge with the right people. His wife was in Chicago most of the time, and he had this friendly, easy temperament. He never rushed anybody, he never made anybody mad, but he exerted lots of influence the slow way, which none of us young people had the skill or imagination to do.

Somebody arranged, if Groves would not let us meet with the Senate Committee on Atomic Energy, the Army could not turn down a dinner at our expense which involved all the Army he wanted to name, plus the Senators. So we all met in the only nice restaurant in Oak Ridge (it must have been the Grove Restaurant, brand new; it was so new I'd never been there before) and talked a little bit, and I met some of the leading people of the other groups. To my horror it was suggested that I be master of ceremonies because I knew two or three people in each of the other two groups and I knew the Oak Ridge people fairly well. My name was known to the chemists in the other groups, and the other laboratories were more chemical than ours, so they felt better with a chemist than a physicist. I wanted Warren Johnson to do the job, because Johnson—Johnson was smart enough to get me to do it.

In walked the senators, and, as a captain opened the door, in walked about 15 colonels and two generals: two-star General Leslie R. Groves (this was the second time I ever talked to him; I think I may have seen him lecture once or twice before) and buck [Brigadier] General Kenneth [D.] Nichols, the cold-blooded fish who was the commanding officer in Oak Ridge—a very competent man. It was all sweetness and light. Everybody shook hands with everybody else, and the sitting arrangement was to be a senator, a scientist, an Army officer, a scientist, a senator, a scientist, an Army officer, a scientist, all around the table. The conversation was friendly and easy, and I found to my great delight that one of the colonels was Bill [William Reeves] Shuler (~1912-2007), who was Caltech's only All-American football star who got a bachelor's degree [engineering] at Caltech in my year [1932] and we won the [conference] championship that year, the only time in history. [1931 and 1930, Loma Karklins, CIT Archives, email 13 July 2010.] Then he went to West Point and was allowed to play one more year, and that's when he made his All-American.

Shuler had just come back from Italy. The Army had loaded up lots of colonels from the Engineers from Italy, who were thrown into Oak Ridge to learn about atomic energy before the Army would lose control when it became civilian. You see, this was inevitable, though the atomic energy act hadn't been passed yet. The atomic energy act, as I recall was passed in early 1946 [signed August 1]. The Senate was trying to find the information.

Shuler told me he was involved with foreign activities in atomic energy. I asked, "How can you know anything about that? You've only been in this game a couple of months." "Well," he said, "we get reports." I said, "What sort of scientists do you have in places like Saudi Arabia and Persia?" "Well," he said, "some oil geologists send in reports, if you want to know." "But," he said warmly, "anytime you come to Washington, you're welcome in my office." I'm sorry I've never gone to his office, I don't know where he is now. But he was very warm to me.

The man on my left was a colonel who was in charge of the administration of the Oak Ridge town site, as I realized after a minute or two. He was the man who had gotten a very hot letter from Grace Mary. She topped Wilson's 14 points by 15 points as to what was wrong with Oak Ridge. She was hot under the collar at the Army control of our life. But he found out I was a civilized guy and I kidded him a little bit about my wife writing him a hot letter.

Things went fine, except that I had, when the dinner was over, to move to the head table. Our program had been planned involving a number of us speaking about various aspects of atomic energy, but Senator McMahon pressed us. He saw the program and he said it was too long. He said, "Remember the Senate is on Washington time. Make this short, make it sweet, and make it effective, and don't let General Groves bulldoze you."

Well, we had a series of talks. One was "Atomic Energy, Peaceful Uses," by Alvin M. Weinberg. One was—Lyle Borst talked on "Atomic Energy, Peaceful Uses in Physics and

Chemistry." He was a chemist who had been in physics. Weinberg was speaking about peaceful uses of atomic power. Paul Henshaw was talking about biology. All the speakers were Oak Ridge X-10 people, because we had the broadest view of atomic energy. The last speaker we had nobody scheduled for, was atomic energy in war, but we had hoped to get Stafford [L.] Warren (1896-1981). (His wife, by the way, was quite interested in Oak Ridge history. He's now head of the UCLA Medical School, but he was a colonel in the Army.) McMahon thought we had too much stuff but he said, "Take that last one and get it good." He said, "Who are you getting?" I said, "We want Stafford [L.] Warren (1896-1981), we haven't had a chance to get hold of him." He'd been with the senators all day; he'd probably come in from Washington. General Groves had these people under tight control.

So I went down to ask Warren—I knew him fairly well—and I said, "Will you give this talk?" Warren sat within hearing distance of Groves, at the tail end of the table. I suppose we should have had Groves at the head table, but we treated the generals no better than the senators or scientists. I, as master of ceremonies, started out at the middle of the table, as had been planned. He said, "I can't give this talk without General Groves' permission." So I turned to General Groves, and said, "General Groves, would you mind if Colonel Warren tells us about the atomic bomb at Hiroshima?" He said, "Of course I'd mind. You can't give that talk." I said, "I don't know why we can't, it's on the program. We ought to give it." He said, "Nonsense! We're telling the Senate this tomorrow. I won't have you have that talk on the program." I said, "But General Groves, we have to have it." He said, "You do not."

I went up and told McMahon. McMahon said, "Get somebody to give that talk. The hell with this general." I said, "All right." He said, "How are you going to do it?" I said, "I don't know. I can't pass a note." The man I wanted to give it was somebody pretty imaginative, and I said, "I'll get Al Weinberg to do it because he seemed to the people on the program to be the one with the most flexibility." But he was sitting directly across from General Groves. He had probably heard the conversation I had with Warren and Groves. I'm not sure he did, and I had to count on this.

Well, the talks began. I decided I had to cut the Army down to our level. I had to take a risk of taking too long a time. I hoped each person would talk about 10 minutes. I think Brien McMahon introduced the senators to us. Then I introduced, first of all, General Groves, and then General Nichols, and then I think no other Army officer because I didn't know very many of them. Then the scientists were to be introduced as they would speak and there was no point. I probably introduced Warren Johnson as my boss, sitting I think right next to me at the head table.

I decided I had to cut the Army down to our size and I would do it boldly. I introduced General Groves by saying that I was supposed to have a fantastic memory but General Groves has an equally good memory, best memory I ever saw in anybody outside of science. I mentioned a few more things. Then I said that General Groves had not only memory, but energy. He was in close contact with all of the contracts for atomic energy. I understand the number was 830. I knew I was roughly in the right ball park, but I knew I couldn't be right exactly and Groves did just what I wanted.

He got up and said, "I'm sorry, sir, the number is 870" or whatever it was. He corrected me. In other words, he brought himself exactly to my level, which is what I wanted.

I introduced General Nichols as the only man known to me in the U. S. Army with an earned PhD. Nichols blushed and blushed and blushed. I didn't realize that in the presence of a commanding general, of a person like Groves, you just don't praise other people, you just made them out sorry. He got up limply and took a nod.

Then I introduced the first speaker, I don't remember who it was, and we had—I think it was a fellow named Stewar[d, t, or Stuart?] from Y-12, who was also speaking about isotope separation. I had pointed out, now here's a man who's assistant professor of chemistry at some college, I'm the same level, he lives less than 100 yards from my house, and I never met him until this night. It shows how busy we've been in Oak Ridge, but we have common interests and I'm sure that from this night forward we'll be good friends. Then Stewar[sp?] gave a talk—I don't remember anything of it.

Then I introduced Al Weinberg as giving a talk about the atomic bomb in war, but I began by saying that Weinberg was a man that came to the Project about the same week as I did, but being in physics he'd been in the game earlier. Indeed, he'd been in biochemistry, a physicist employed in biochemistry such as studying the problems of the fusion of metabolites in cells, and it turns out these problems are very much like the diffusion of neutrons in the reactors, so he not only had a background in physics, but he'd been in the laws of neutron physics before neutron physics was realized. This is not necessarily exactly the story I told, but I told the story in this spirit. At any rate, I kidded about the thing. And, I said, Dr. Weinberg has all this imagination and all, he's going to tell you about the atomic bomb in war.

Weinberg got up and looked puzzled, and went on and launched a speech about the atomic power in peace. He talked about atomic power plants at the North Pole, and atomic power plants—well, you have to have them away from cities in the open spaces and all, and he talked about fused metals and fused salts, and all the trash that we hear about all the time. He went on, and on, and on. He probably talked not 10 minutes, but 18.

McMahon knew I'd missed my boat, and he said, "Stop the man!" I said, "I can't stop the man with anything short of a rifle. He's sitting right across from the general, and I can't stop the talk." He said, "What are you going to do about the speech?" I said to McMahon, "I'm going to give it, as a rebuttal to this."

Finally, Weinberg sat down but I'd taken a few notes of the high spots of his speech, and I got up and I said, "Well, I told you I've know Weinberg a long time. I didn't tell you we are always on the opposite sides of every arguable question. Dr. Weinberg has told you about atomic power in peace, but I would like to tell you about atomic bomb in war, because that's the thing you're going to realize. It isn't going to be the empty spaces of the world, it's going to be the crowded cities of New York, Washington, St. Louis, Los Angeles. It isn't going to be liquid metals carrying the heat away, it's going to be vaporized metals like the tower of Alamogordo."

So I gave the speech. Everything was an inversion of something Weinberg had said. Now Weinberg caught on what was going on. General Groves knew what was going on, and he had the blackest face I've ever seen on a man in my life. McMahon was enjoying no end how I'd finally gotten the thing he thought would never come. I didn't have the nerve to say that Warren was forbidden to speak and I was going to give it. The only people who knew what was going on, really, were Stafford Warren, Groves, and McMahon.

Then Borst gave a talk which would have been good in any other context, but it was anti-climactic. The meeting lasted till about nearly 11, which is nearly midnight Washington time. Before the dinner all of the Army had been warm to me; after the meeting no Army officer came within 50 feet of me. Every senator came by and thanked me, on behalf of the scientists, for a most illuminating evening. I walked out a dishrag, went down to the guesthouse in Oak Ridge with somebody.

Gosh, I didn't want to talk. I did want to tell a few people what had gone on, but I don't remember who they were. Weinberg must have been one. I went in the guesthouse to get some cigarettes, and General Groves walked in and didn't see me. I had my back to the desk. He said, "That's what you get for free speech!" Then I went home and told the story to Grace Mary, and I lay on the floor and talked till about 3 a.m. just to unwind, because this was the tightest fix I'd ever been in, in my life.

JBS: That's a magnificent story.

Coryell: That's a good day's work. You want some more stuff now?

JBS: Well, yes.

Sometime in the fall of 1943, I realized there were some people on the Project I just worshipped. These were people who were, without exception, able scientists and they were also, without exception, men of political feel and courage. I also tended to compare these people with two different scientists known to me before the war. The three leading scientists known to me before the war intimately, were Arthur A. Noyes, Linus Pauling, and I'm not sure I would have thought of Franck in this sense or not. I had two very illuminating days with Franck at UCLA before the war, but Franck got in on the war.

I decided that I had a circle of gods, and there was no question in my mind that Noyes would have been president had he been living, but he was president emeritus since he was dead. He died in June of 1936. James Franck, without any question, was president. I talked about this with a few people, and the number of these gods was fairly high. Pauling was a vice-president—he was a notch or two below Franck. He wasn't as courageous as Franck. What I loved about Franck was that in spite of ill health, in spite of a thick accent, his knowledge he was a Jew and ought to keep out of things, he never stayed out of any fight and he didn't care whether he was going to win or lose it. He spoke what he thought was right, and he would fight vigorously and bitterly, but he didn't waste his energy (like I do) in blowing off steam.

I felt that Samuel K. Allison was one of these gods. Eugene Wigner was one of these gods. I'm not sure whether I considered Szilard one or not. The funny thing is, I almost always think of a man and the answer is obviously yes or not. I think of the time a man got elevated to a god. Karl Compton was elevated to a god, not the first day I met him and not because of the telegram or not because of the action of his about the cyclotron, but shortly after I'd seen him a couple more times at MIT. Arthur Holly Compton never made the grade. I used to joke that there were new elections coming, and there'd be nominees, but I never mentioned a nominee who wasn't going to be elected. I think Szilard is certainly a god now, but I guess he wasn't a god in those days. He was in an odd position in the Project and this was before I saw his magnificent underground work in Washington. It's obvious he's a god. It's just that he wasn't a god in 1943. Matter of fact, I got to know him much better in 1944, and through this stuff in 1945. Have I mentioned J. Robert Oppenheimer?

JBS: No, you haven't.

Coryell: Well, J. Robert Oppenheimer I knew slightly, and he was nominated to godship from a slender contact and the fact that everybody at Los Alamos felt he was a chief god. He obviously was of god calibre. Oppenheimer is certainly a vice-president of the gods, and I knew this already in 1945. I'll come in the next time with what I had to say with Oppenheimer that very evening at the science end of the dinner, and why I think he did like he did about the May-Johnson bill.

Fermi is there, but he's the one exception. Fermi is the most brilliant man I've ever seen and I saw no evidence of a conscience in Fermi until 1954, in the Oppenheimer case. The first political thing I know about that Fermi ever did was sign a protest about Pauling's being denied a passport. The second political thing I'd ever known him to do was try to persuade Teller to make peace with Oppenheimer the day Fermi went into the hospital for the exploratory operation shortly before [blank]. [Fermi died on November 28, 1954.]

Sugarman was always jealous he didn't have any godships, but Sugarman has the unique position of archangel in this section of gods. I think I told you the number was 13, does it add up?

I guess being a god doesn't carry any honor. I haven't talked about it very recently, but in '45-'48, I would be proud to say, when somebody would say he was going to Chicago, "You go to see Franck. He's one of my gods. You go to see Sam Allison. He's one of my gods."

Well, that's the list of gods. There hasn't been a meeting of the committee of one. I haven't considered, analytically, the situation of gods since 1950, so if I reflect back it's possible I would find that if Szilard wasn't, he is. You see, this is an example. In essence, Eugene Rabinowitch was in that category. I'd certainly label him as one of the gods, and Victor Weisskopf. Victor Weisskopf I didn't know until after the war. I heard someone from Los Alamos say he was a sort of a bright guy but not a heavyweight, but there's no question in my mind he's a vice president.

There's a probability if I'd known Don[ald James] Hughes (1915-1960) a little better he'd be in that category. It's possible, if I reflect over people I've known in the last few years, there'd be one or two more candidates. George Scatchard (1892-1973) at MIT is close to being—he's just elected.

Session Eight — 19 August 1960

Coryell: We were talking about 1945, October 25th.

JBS: Yes. What I wanted to ask you was, at this dinner of the senators and scientists in Washington on October 25th, you mentioned that Robert Oppenheimer's position had become clearer to you in regard to the question of the control of atomic energy after the war.

Coryell: Yes. This is the result, of course, of a friendly contact with Oppenheimer. I was probably the first Oak Ridge resident to meet Oppenheimer after the war. He came to Oak Ridge rarely during the war, and I suppose he didn't make a trip for some months afterwards, though prominent physicists might visit Oak Ridge and also see Oppenheimer in Washington.

Nobody visited Los Alamos without being captured and staying there. I might say a few words about Los Alamos. It was all a one-way street of information from all the projects to Los Alamos. Rarely did anybody at Los Alamos have the imagination to send information out to the sticks that would help us do better work for Los Alamos, but there was also no resentment in the war. The same principle still exists to some extent in Los Alamos, though it isn't a wall of secrecy.

When Seaborg visited Los Alamos in February 1943, shortly after I had the fight when I told him all the people I knew that didn't like the way he ran his group (I may have told this already), he said he'd been to Los Alamos. I also think I told that Teller said never to worry about Seaborg at Los Alamos, that Oppenheimer had told the guards to shoot at the sight of the whites of his eyes. But what happened was that Seaborg came to Santa Fe, to the La Fonda Hotel, and phoned Oppenheimer, and Oppenheimer said, "I'll come down and see you," which is what happened. And Seaborg didn't see the inside of Los Alamos until roughly a year after the end of the war. But all of us in Chicago assumed he had, because he wanted us to assume he had.

There was a unity of spirit at Los Alamos, as you know from your own childhood there. Everybody knew each other well, the bickerings and all were very small. There was a tremendous community of spirit. I think that the ALAS, the Association of Los Alamos Scientists, was the most vigorous of all the spontaneous groups of scientists, and as far as I know it was nearly total in its coverage in the fall of 1945.

Well, here I was, a babe in the woods in Washington, with guys like Shapley, whom I had never met before and Szilard, whom I knew very well and loved, glad to have a scientist to speak for Oak Ridge, meeting Oppenheimer and the people from Chicago who were there. I had breakfast with Shapley and he invited Henry Wallace, the former Vice-President and Secretary of Commerce to the table with us. Senator [Arthur Hendrick] Vandenberg (1884-1951) was at a nearby table. And I think I had breakfast one of these days with Oppenheimer. If I find this memo I mentioned, the odyssey of Charles Coryell, these things will be spelled out with exact dates and the like. I also spoke to Oppenheimer at the cocktail party before this evening's talk.

I asked him rather frankly how he could be for the May-Johnson bill, and I think I drew to his attention the telegram from Karl Compton. "Charles," he said, "there's no question of the urgency of civil control of atomic energy. But far more important is the international control of atomic energy. The important thing is to get any sort of an act through Congress in a hurry and not worry about it. I assure you, Groves will never be Atomic Energy Commissioner."

I said, "Isn't it in the act?" He said, "Never mind." And then I found out a minute later he was on his way to an appointment with Mrs. [Eleanor] Roosevelt (1884-1962), and one of the other people of the high-ups told me he could see President Truman anytime he liked. He was essentially Mr. Atomic Energy, and what he thought made sense about personnel would take place, but he couldn't bulldoze foreign affairs until we had a law that had something in it, and that was why this law was good enough and an internal fight about the law would be utterly divisive.

When I brought this news back to the scientists in Oak Ridge, this helped relieve some tension, because people had been very uneasy. There were already in the fall of 1945 meetings in Chicago, or in Washington, or in New York, where some scientist would come in and speak with some people prominent in public affairs—Raymond Gram Swing, Norman Cousins (1912-1990), men of this class. There had also been some negotiations should these spontaneous chapters of associations of scientists federate. Indeed, in the fall of 1945 there was a fantastic fight whether it be called

FAtS or FAmS—whether it be an association of atomic scientists exclusive to people who worked in nuclear science, or whether it should be a federation of American scientists (which it turned out to be) with atomic scientists in a prominent position, but it could be diluted out with a bunch of biologists or oceanographers or astronomers or anyone else that wanted to capture it. It turns out nobody wants to; the Federation of American Scientists is required by its charter to be two-thirds people who are professional scientists—bachelor's degree or equivalent—and there's never been any trouble about this. Out of its membership today, physics is a prominent group and 80% of those in physics are Project alumni or working in daughter fields from the atomic energy project.

When meetings were held, Chicago would send a delegate, Oak Ridge a delegate, Los Alamos a delegate, the Association of Cambridge Scientists a delegate, and then there were four or five other recognizable centers I mentioned—in the Allegany group in Cumberland, Maryland, and other California centers. They didn't always have a scientist travelling East who could represent them, but we had a very liberal plan. If any scientist had business nearby, he would be a spokesman for the others. He would get a quick briefing the night before he left.

In these negotiations, the Los Alamos people would generally go along with the rest of the people. Then they would send a telegram, or telephone back to Los Alamos, and they would find that the rest of the Los Alamos people would kick. The Los Alamos people wanted total support of Oppenheimer's position. This is the way it looked always from the outside. And the Los Alamos Association executive committee would recall the delegate, just like the Russians do all the time. We actually said that dealing with Los Alamos was like dealing with the Russians, but we didn't realize the total confidence they had in Oppenheimer and we also didn't realize that Oppenheimer really had the inside track, that this was justifiable.

When I brought the news to Oak Ridge and to Chicago, this also helped lower the tension of the situation. It also gave me a chance, even though I saw Oppenheimer just this evening and at breakfast and a couple of other times—and later on I heard him give a wonderful lecture at Oak Ridge, just contorting himself thinking this out. I saw this speech from a wire recording, and I saw the difference between the written speech and the spoken speech. The speech was utterly moving, and reading it you could see the pain of the English. Later on I saw it in some publication, smoothed out, and it was again good English, as Oppenheimer is capable of writing. I got a little feeling for his problem. There was no question in my mind that Oppenheimer belonged in the circle of gods.

I was in the Oak Ridge group some time before the senators and Army officers came in November. The groups at X-10 and Y-12 and K-25 amalgamated and one of our big functions was local education. Matter of fact, I think it was probably in the fall of 1946 that we got some money from the [William] Volker Foundation (1932-1965) and sent a trailer with an exhibit and two or three scientists across Kansas, talking to high schools and towns. Lyle Borst was in charge of this. There was a lot of activity in public education.

On Mother's Day in 1946, the first anniversary of V-E day, I gave my first public speech. Our next-door neighbor, Lib Taylor, was active in a church in Kingsport, Tennessee, and she invited me up to talk to her church group in the morning and give a public address in the afternoon in the high school auditorium to the League of Women Voters. It was the first of a long and fruitful contact with the League. I was scared to death. The Sunday school group didn't bother me at all. I used to be a Sunday school superintendent, in the Mission of the Holy Spirit [Episcopal], in Wilmar, California. I was speaking to young people, so I talked sort of just the importance of science, the dignity of science, and the obligation of worrying about the applications of science. I spoke how I felt without worrying how it would look in the wire report—it wasn't reported.

In the afternoon I felt more obligation. I was being introduced by a minister, they were adults and all, and I had an outline of about two-thirds of a page—the topics that I'd cover. I started out painfully, I'd got tucked away and I'd forgotten about the thing, and I talked just what I was thinking at the time. As always, I began with a little picture of what uranium was like, and plutonium. You had to tell this story; you still have to tell it. The public doesn't know what isotopes are. I'd give a little bit of that and then talk about the promise and the threat of atomic energy—the threat [blank], and the fact that public education was important and public reaction was important. And, of course, there was always the drama of the fight between the May-Johnson bill proponents and the McMahon bill, and there's always the necessity of adding positive support for McMahon's liberal position, and to project liberal action by the United States and the United Nations.

Not long after I spoke for the newly formed League of Women Voters in Oak Ridge, I spoke at the Oak Ridge High School. My first year at MIT, I gave 65 outside speeches. I donned my doctor's robes without the hood and cap and gave a sermon at the Centenary Methodist Church in Auburndale, Mass., on World Peace Sunday, which must have been about October 20th, 1946. The minister knew I was an atheist. The minister liked me very much. I had written him a letter in response to a speech I heard of his over an arbitrary radio program at 7 o'clock one morning. He said what I thought were very significant things about the human side of atomic energy. He asked me to call on him and I did. We talked, and I said, "You don't mind an atheist giving a sermon in church?" He said, "You're not an atheist by my standards. I assume you have enough good taste not to antagonize my congregation."

I sought a Bible quotation from St. James about peacemakers, (I can give this when the tape comes back— James [Matthew 5:9] something-or-other, "Blessed [are] the peacemakers"). [The Letter of James though short, reveals a telling influence on CDC.] And I talked about the moral problems of atomic energy. For many years I began almost all my speeches with the following statement: That when I was a boy, I decided I would never be a diplomat or a soldier, because they are responsible for the mess of the world, and war. I thought being a scientist would be a very noble thing. Before that I wanted to be first a missionary, then a medical missionary or an agricultural missionary. I guess I have a Messiah complex. But now I'm a scientist and we're war-makers, too, but scientists more than diplomats and soldiers are concerned about this. At least with part of our mind we're trying to work for a peaceful world. And though part of us, including me, still work on things like atomic submarines. If I were asked a question on bomb technology by a government employee, and I knew the technical answer, I would tell the government authority.

There's a story about Norbert Wiener (1894-1964) though, who made calculations for the Air Force during the war at MIT, and at the end of the war he was asked for these calculations—about the spring of 1946. He wrote an elegant letter to the Air Force denying them this information. He said, "Of course you can get it in the carbon copies that were distributed during the war, but I refuse to be responsible for passing this information on." It made a nice letter in *Science,* it made a nice newspaper program, but this isn't the way I feel. I've always, since the end of the war, had an appreciable income—15% of my income is consulting to atomic energy laboratories. With two exceptions this had nothing to do with anything military except the background science of fission. The income is welcome, but more important, the contacts with what goes on scientifically in these fields is very welcome. My own office has been an information office for non-classified science and my friends behind the classified curtain have given me non-classified leads and can tell me when I should ask for a classified document which I still have the right to see.

I'm a man whose Un-American anti-integrity is guaranteed because I have two clearances which have been continuous from the start which proves I cannot be a Communist. I never have, and I don't intend to be, but I have a government certificate, in a sense. I carry in my pocket a Los Alamos clearance, the only one I took to Europe in 1953, which I could use if I had to go to an embassy to explain that I had to make a report and that it was all right for them to talk to me and I would make some sort of a report.

I'd like to tell several security stories, though, before I get too far, since we've sort of come into this problem of scientific information.

One of these happened in the fall of 1944, or the early spring. I don't remember now when the Hanford plant first went critical, but my impression is it was November, 1944. Now, Hanford recruited people out of the Clinton Laboratory personnel, out of the Metallurgical Laboratory of Chicago. It sometimes recruited with difficulty. I would have gone had I been approached for a position where I had freedom, but I didn't expect freedom from the DuPont Company and I think some people in the DuPont Company wouldn't want to give me the freedom I insisted on, because I felt that the work in Hanford was important. Heelan [?] refused, as I've told, and when Nathan E. Ballou was taken from Heelan's laboratory, he went. He felt it was an interesting thing to do, but more important, he felt it was necessary and he would do it. The boys put a service star in the window. We had one of our boys away in the services.

Truman P. Kohman in Seaborg's group also went. Another was William H Sullivan. (The H has no dot. Sullivan's middle name is H, and he has no initial. Different from Harry S. Truman, whose middle name is S. and has an initial, S.) [Cf. Trumanlibrary.org/speriod.htm.] Sullivan had gotten his PhD just before the war started, roughly 1940, with Kasimir Fajans, at Michigan. He was one of the great pre-war nuclear chemists, and one of those being considered to replace Latimer, but had been turned down because he was an older man and felt that no radiochemist was trustworthy whom he hadn't trained personally for a year. Allison wouldn't tolerate training Coryell and Boyd and Burton for a year. We were working the first week we arrived.

I first met Fajans personally when he came to that Chicago interview, though I had hoped to work with him when I went to Munich. (I think I told this in my first talk.) Sullivan was one of several radiochemists from Fajan's laboratory whom Spedding collected in his group at Ames, who tried to do a pretty broad front in radiochemistry, including plutonium chemistry which angered Seaborg and which partly explains Seaborg's disgracing of Spedding to weaken his group to save worrying about the competition.

Another man was Adolf [F.] Voigt, who—I won't try to give more names right now, because these two I saw the most of.

I was personally very devoted to Spedding in these early years, and went to Ames three or four times with Boyd. We felt it was a good group, but sort of cut off from the mainstream. Spedding tried to protect them from the hurly-burly of Chicago, he thought they'd work better, and he was their only spokesman. Except for occasional visits like that of mine and Boyd they were pretty much cut off from the excitement of the Project, and also from the training one gets from this fantastic confusion in Chicago.

Sullivan was glad to leave Spedding's tight control and come to Oak Ridge to join my group in February 1943, if I remember the date right. Then he was willing to go on to

Washington State, that is, Hanford. The chief radiochemist at Hanford was John E. Willard, whom Seaborg had brought in at about the same time I did, and whom I felt was a tool of Seaborg in the early days of the Seaborg fight—particularly in the ion exchange aspects—but I've since realized was just doing the best he could and didn't know of the overtones of the conflict with Boyd that was going on. I know Willard very well now, and I know he's totally honest and very capable. Also he treated at Hanford the men who went with him quite decently.

But the DuPont Company was anti-research. They only wanted the minimal research to protect the company in case of trouble and a lot of service work to make production come fast and quick. I felt I had a moral right to communicate with Sullivan at Hanford comparable with the one I felt I had to communicate with Sugarman at Chicago, which had been guaranteed by this sort of treaty that Franck had arranged for. But I had no such legal right. Sullivan was not my parallel at Hanford, Willard was. But Sullivan was my close personal friend, and early in the game we outlined a code in a classified letter so that we could talk on the telephone about fission products. I felt it was appropriate to talk whenever we could get a chance, but he wanted to communicate with me by letter. He had to send a letter signed by the director of the laboratory and I had to respond the same way.

Well, he was glad to talk over the phone. Lots of us became "phony" engineers, or "phony" scientists. He called me a couple of times. About a month after the reactor started, but after the sleeping sickness problem I talked about some time ago had been resolved, for some reason he phoned me. I asked him some questions. I was simply enthusiastic. I could talk about buildings and operations at Hanford in terms of the Oak Ridge code. The reactor at Oak Ridge was the 105 building. Probably they had the same number out there, because DuPont had a code about buildings where 1 indicated manufacturing, and 2 something else, and 7 was research. So I could talk about operations this way, and I could find out that the reactor was working pretty well, and that it was two-thirds power. These things were done in a way that was probably not a security violation in telephone calls, but we were under strict rules not to talk about things in case the line were tapped.

Then I asked him about some matters of how much freedom of action they had, and why he couldn't do some experiments that he and I had talked about before he left that would be of long-term value. An example of this, though I'm not sure it was a subject of conversation, was a job that Johnny [John T.] Swarthout, now associate director at Oak Ridge, did I think with Sullivan, to find evidence that the nuclide Rh^{102} [rhodium-102]—they settled, not preliminarily, the amount formed in fission. The DuPont Company thought that this was trivial. There was not enough of this to be of any danger to the plutonium process, but if one wants to deduce how the fission process works, what the properties of fission are, something still not very well known, one has to know facts like this. Sullivan and Swarthout's number of fission yield of less than 10^{-6} percent is still decisive in theoretical interpretation of fission.

Now these men were keen to do these things and, if they presented it right to the DuPont Company, they could. But Sullivan had had some bad luck in authorization that particular week, and I started asking him, "What wouldn't those bastards let you do?" or "those dirty, stinking bastards." We used the word bastard repeatedly. It was also a code word for the element zirconium. Super-bastard was a code word for the element niobium. These are the elements 40 and 41, which are awfully tough to handle chemically. So the word "bastard" applied to the DuPont Company came frequently, and the word "bastard" in these other two usages only, but any chemist could tell which was being talked about. We must have talked about 40 minutes and I ended up feeling sorry for Sullivan but glad we'd had a talk and that we'd both gotten some feelings off our chest.

About three days later a personal letter came to the house which Grace Mary opened. She always opens my mail and I always open hers, this is no hazard. It was from Sullivan, mailed outside of the Hanford project—from Pasco, probably. It said, "Charles, our goose is cooked. Our phone call was transcribed and copies have been sent to the director of the laboratory." Sullivan was in great anguish. He felt sure we would both be ejected from the Project and humiliated and thrown out to the wolves. But I broke out laughing. I knew we'd talked so sharply about the DuPont Company, and that a fair fraction of the things we said were so true, that it would be embarrassing to the Company. The DuPont Company might have trouble destroying the record of this phone conversation, but any intelligent DuPont Company officer would be glad to, because they would never want it to get out into any non-DuPont circles. It was plenty of ammunition for any anti-DuPont people, and there was plenty of anti-DuPont feeling everywhere in the Project and there probably was in the Army.

Sullivan apparently got a lecture from the security officer at Hanford (I'm sure this happened) but probably not a very serious lecture. I suspect that the lecture was not on the grounds of bad taste, but about the possibility of a security slip. But Sullivan and I did have the protection that our chemistry was pretty well clouded in bastard and a lot of other stuff and a non-chemist would have trouble figuring it out. I never heard from the security officer at Oak Ridge. But I was just amused, not frightened, because I figured our goose isn't cooked. The DuPont Company has just gotten, unexpectedly, a very sharp reprimand for the poor way they handle intelligent scientific personnel.

I'd like to tell another security story, which is now going to move ahead for some time. It's now 1951. In 1949 I was delighted to have my first foreign scholar, who wanted to come to MIT to work—a young Norwegian citizen, born in London of Greek parents, Alexis [Georg] Pappas (1915-2010). He had gotten his bachelor's degree, I think in radioactivity, with Ellen Gleditsch (1879-1968), Norway's then most prominent nuclear chemist, an old lady. I think she's still living. Pappas had, during the war, worked in an acetylene factory and been involved in underground activities. After the war he had a fellowship for two years to work with Moïse Haïssinsky, a slightly left wing, very able, French radiochemist with Italian training and Russian-Jewish origin. He'd also been in Israel as a kibbutznik, in 1930, he'd been one of the chemists with Fermi, and he'd worked with Marie Curie.

Pappas wanted to do a big enough job in some university that he could write a thesis in English for submission to the University of Oslo. He could have taken a PhD at MIT, except that the rules are so strict about things like organic chemistry that hardly seemed useful. Furthermore, a PhD in Norway comes to an able man for a major body of work, not a major work, but a body of work, when he's about 35. MIT would have given him a PhD but he'd have less credit. It was to his advantage to have one in Oslo, as well as to my advantage. He came in September 1949, and left in October of 1951, with a few weeks at Brookhaven on the way out. About a year after he left MIT we put out a 180-page MIT report which MIT printed and which, with a separate cover page, was his doctoral thesis in Norway. But he was so exhausted by then that he never published a book, which is traditional for a candidate for a degree, though he made several important papers for the Geneva conference in 1955 out of this.

Pappas wanted to study fission yields in the mass region 129 to 140, radiochemically. He did a superb job disentangling fission yields. We also had in the laboratory a young Canadian, Donald R. Wiles (1925-), who had gotten a master's with [Henry George] Harry G. Thode (1910-1997), at [McMaster] University on fission yields by mass spectroscopy. Wiles was fairly imaginative, and he had seen evidence that nuclear shells affected the fission process. Pappas wanted to probe this further. Pappas, of course, did a much bigger job than Wiles. He was a mature chemist and a very likable person, and we talked together a lot.

But in the background of this was the fact that mass pattern 135, which was right in the middle of the Pappas pattern, right in the middle of the Wiles pattern, has as its chain 6.7-[hour half life] I^{135} [iodine with mass 135], decaying to 9.2-[hour half life] Xe^{135} [xenon with mass 135]. This xenon-135 is the poison which causes sleeping sickness in reactors and its neutron properties at that time were supposed to be secret. Wiles knew about this. It was known in Canada, and it was known virtually all over the United States. He was empowered to speak with Thode, as a result of some earlier arrangements made with Canada, but I couldn't speak to him about this and I never did. ["The xenon-135 situation led to the fact that Charles had a copy of my M. Sc. thesis in his safe, but I was not allowed to know that. One day he actually showed me the cover of the document, but put it back into the safe." Don Wiles, email, 28 August 2010.]

Pappas was a "furriner" who had no basis to know these things, and never am I aware of even giving him a hint that anyone in the Russian government or the United States government or anybody in the atomic bomb or atomic power game was interested in xenon-135. But I was delighted he wanted to work in that area, and nothing he did would give neutron properties. But the fission yields he was after were important in reactor design and, I think, to a very limited degree in bomb technology but in the getting of fuels for bombs. I think Wiles may have given Pappas some information. This is not my problem and I have no knowledge of it. I never gave any information to Wiles, but Wiles and I had an understanding based on a few words that I wouldn't tell him anything about this, that he knew that something was very interesting, and I couldn't tell Pappas anything about it. And there wasn't any need to tell Pappas.

In February 1951, I was asked to be a consultant to the Westinghouse atomic power division at Bettis Field in Pittsburgh, through the interest of Ronald A. Brightsen, who took a master's with me in the fall of 1950. They were interested in designing submarines, and submarines are very sensitive to sleeping sickness because they have to be maneuverable at all times, irrespective of how much maneuvering or lack of maneuvering has happened. The answer is often to make two motors or more in a submarine. The design problems of the [USS] *Nautilus*, which specifically we were talking about, were based on data gotten by Sugarman and partly by my former people at Chicago, and independently by [blank] [Bill Walters suggests this may have been Art Wahl centered at Los Alamos, email, 29 August 2010] and me personally at Oak Ridge. We had been in competition and our answers had zeroed in on a number for the fission yield of this chain of around 5.9%. Pappas' data showed rather soon that the real answer was about 6.4%, and I happened to mention this in classified circles about June, 1951.

This caused a catastrophic fright. It was brought to the attention of high officials in the Navy, and undoubtedly to Admiral [Hyman G.] Rickover (1900-1986), a man I'd met a couple of times. It was decided there should be an emergency meeting on this. One can argue that the principal reason for this meeting is the Navy had to predict the operation of a submarine. If it wasn't well designed, and it failed, at least this should be predicted so the Navy couldn't be blamed. But if possible, the design should be changed to take care of the fission yield if the Pappas figures were right.

I spent the summer of 1951, or at least six weeks of it, at Los Alamos and talked about this problem there. Then, at the time of the meeting of the International Union of Pure and Applied Chemistry in September, 1951, in New York, and also at the meeting of the American Chemical Society (I think it was the 75th anniversary). We planned to have a meeting of Project specialists. This was arranged by me, by Lionel Goldring at Brookhaven, a former student of mine, who knew the classified side of the problem, and by Brightsen, who knew the classified side of the problem. Among the invitees were Donald R. Wiles, a man not clearable by the United States because Canada wouldn't designate him as a major scientist (he was only a graduate student) though in Canada he was cleared; and Alexis [G.] Pappas, who was not clearable by any of the then standards unless his work should be proved to be of tremendous importance, and nobody wanted to guarantee this.

The meeting was held in the AEC building in Columbus Circle in New York, about two in the afternoon. I think we had lunch at the expense of Dr. V[oscan L.] Parsegian. He asked me to open the meeting. He was a sort of security officer; he was representing the AEC in the New York office. There was Paul [F.] Gast (1917?-1974) from Hanford, and a lot of bright people. There were some bright people who knew the reactor side of the story, since you could measure this by reactor physics.

Parsegian and I had talked that we would have a general discussion not involving Pappas, but have Pappas sitting in his office to be on call in case we needed some facts that I couldn't remember since I hadn't been at MIT in three months and Pappas and I hadn't looked at the data—he had just told me. But Pappas by then knew the fission yield of this chain was important and I'm sure he knew what it was. I said I would talk in generalities and we would have a meeting for about an hour and not disclose any classified numbers. Then Wiles would withdraw and we would then

328

talk about the ultimate problem, involving a liaison with Canada not for Wiles' sake but for Chalk River's sake.

So I turned to the blackboard and I said, "Let's put down the facts of the chain. Obviously, some of these facts are classified and I don't want to play around with x. There are going to be too many unknowns in this problem anyway. So I'm going to ask Don Wiles, who is not cleared, to tell me what number he wants to say for the captured cross section of the element [Xe]-135." [Xe-135 specified by L. S. Goldring.] Wiles gave a number that was approximately right, and good enough for the purposes of the argument. Then I put down some of the half lives in the adjacent chains which are involved in the validity of Pappas' fission yield argument.

I looked up and Wiles was gone. I stopped, and asked, and Parsegian said, "I felt that we had better ask him to leave. This is too technical." Parsegian had chickened out before we'd gotten the maximum information out of Wiles. Well, we went ahead and had the meeting, and Brightsen and Goldring together made a first-class memo, which was classified. ["A meeting of AEC, Army, Navy, and Air Force was held in New York in September 1951. He tried to get me cleared for the meeting, but it didn't work. We all started the meeting, he asked me if I knew the (secret) information and I indicated that I did. However, shortly after that, the Security Officer told me that I should wait outside. I stayed outside for a couple of hours while they discussed my own thesis! Those were interesting days!" Wiles, email, 28 August 2010.]

I could then tell Pappas that his experiments were related to a classified problem, and he had the pleasure of helping the

United States, but that it would be a long time before he knew what the score was. But it made me disgusted, the complications we had to do to talk frankly to a scientist about the constructive work he did. I can say that we contributed substantially to the [USS] *Nautilus*—it turns out that Pappas' fission yield is almost exactly right, as confirmed by later work. It also turns out the *Nautilus* didn't have to be redesigned. The normal safety factors took care of this but knowing this fact made everybody happier, so I take credit for the MIT research which was supported by the AEC, but part of Pappas' money came from the Royal Norwegian Research Council and saved $50,000 to $100,000 of AEC business and only cost about $500 to $1,000 for this little conference.

We had another problem when Pappas' thesis came to be prepared. I have authority as a security officer for the nuclear chemistry work at MIT. I'm a recognized security observer, because I know the ins and outs of this thing and I have in my classified files the security rules, which don't give you much information. I've always been very careful on security matters, pointing out sometimes when they've been too lenient with me, but also kicking when they've been too strict. I've generally won, after a suitable number of letters and time, anything I want to be declassified that I think is appropriate to be declassified. I think this is the way you ought to play this game—be careful and honest, and explain why.

I offered to have Pappas' thesis draft—I said it wouldn't be prepared as a secret thesis but it wouldn't be distributed. We had a cleared secretary, but not all the students were cleared—though most of the American students were cleared. I have an open office for visiting foreigners. Pappas thesis draft was prepared this way and submitted to the AEC for review. It took an unconscionably long time for review. I pressed, in letters and on the telephone, for review and finally we got the thing back, but it was obvious that the man who reviewed this had to find several things wrong with it to justify reading 187 pages of single-space typing.

He kicked that we took an average value of the number of neutrons for fission, ν, *nu*, and insisted we carry an unknown x, or carry a range 2.0 to 3.0. And he insisted we find a few other things to change. I think this was just a natural reaction; he had to find something to kick about. I fought this thing to be not larger than an arbitrary 2.5, and got a compromise where I had to insert into the typed thesis a sub-footnote saying, "This value is not chosen with any indication but is a totally arbitrary choice." And in another case I think I had to cite another value we wanted roughly to take the pre-war choice. Because there is the feeling that if a man who knows the answer guesses the answer in what is claimed to be an honest way, this answer is still revealing. This is an argument I can't follow, because it depends on he knows that you know that he knows that you know that he knows that you know this is the number, and you can't tell whether it's an even or odd number of inversions in the story.

Anyway, the Pappas thesis was published, 1200 copies by MIT and the proper number by the University of Oslo. It was spread widely and it received a good deal of attention. Matter of fact, Pappas didn't need to publish his book because in a sense it was a book. It was also published by the AEC as an unclassified document for $1.50, in some thousands, and is now out of print. Pappas did a fine job, but only by accident did we let it help the United States, and only with maximum difficulty did it help the United States.

I've cited this as an example where a man that had data of value to the United States was still excluded from a meeting, and another man who had not only data but judgment from his Canadian contacts was also excluded from the meeting. The Wiles kickout was worse, because Wiles had full discussion of this in Canada.

Session Nine — 4 September 1960

Coryell: Last Thursday Grace Mary pointed out a notice in *The New York Times* that Martin D. Whitaker, president of Lehigh University and former director of the Clinton Laboratories at Oak Ridge, had died of lung cancer at the age of 58. There is an excellent picture of him in the article.

This recalled to mind several things about Whitaker that might be useful on the tape. Whitaker was, at the time of forming the Manhattan Project, assistant professor of physics at New York University and acting director of the physics department. I believe he was at the downtown branch, which has less status than the Heights branch. But many of the people at New York University did research at Columbia, and there were close friendships with Szilard. I think that Bernie Feld came from New York University and worked with Szilard. There were good men, whose abilities and dedication were known to the Fermi crowd, at NYU, and so in the recruiting of physicists these were heavily leaned on. Already on this tape I have recorded the fact that Milton Burton was taken from there. He was then only an instructor in chemistry, because he'd been in industry for some time.

Norman Hilberry, who is presently director of the Argonne National Laboratory, was assistant professor, also. Another old-timer from New York University was Edgar J. Murphy, who was the Army man at the Oak Ridge National Laboratory whom I had a very friendly relation with but often had to accept censure from, in the alarms and excursions I mentioned four lectures ago.

Whitaker was the most dignified of these and was already marked, when I got to know him in the fall of 1943, as the man who was going to head the Clinton Laboratories. He was not a gifted man, but he was a very dignified, decent person, and he knew his limitations and commanded the personal respect of the whole group. Most of us from Chicago were contemptuous of administrators, but we were glad to have one of our own—a physicist or chemist who had administrative talent. I think Whitaker had this.

Whitaker found himself in a very rough position at Oak Ridge. He was ground between the millstones of the rebellious scientists from Chicago and the DuPont Company's high pressure with tremendously able people— the Crawford Greenewalt style and the Army's pressure machine. The Army always operated its pressure machine in line organizations from the top. He was the man who had to be Martin D. Whitaker, director of Clinton Laboratories, above whose signature all of us guys had to carry on our fights when they had to go through letter channels.

I think the worst I can say about Martin D. Whitaker is that he didn't realize the emotional power he could get from us rebellious underlings if he would tell us what was happening. He was always afraid that we would make more messes and make life more difficult for the Army, and therefore impede the Project. He never, in any way I've shown, took personal umbrage to the hostilities that were fought in his office and through his correspondence. I know many things that we did embarrassed him, and he must have gotten holy hell from the Army for some, but he never passed this on or wept on the shoulder of even his closest friend in his most open moment. He was a rather cold, quiet person, and I can think of no one who had a more heartbreaking job than he had.

I often had to fight with him, and he liked me and we could talk pretty frankly when we were alone. I'm sure that this same thing was true of Al Weinberg and a whole lot of younger, emotional people. But apart from the laboratory fights that were going on he was extraordinarily decent, he and his wife, to the people he had time to see. He had limited time. Grace Mary and Julie and I were people that the Whitakers did lots of nice things for. On the personal side I feel warmly toward him, and Grace Mary far more so.

Small things are that we first came to Oak Ridge, Julie was due Christmas Day. The hospital was opening December 10th and it kept being delayed and delayed and delayed, the final delay being for lack of silverware. They finally got a bunch of stuff from the Navy. The hospital opened December 23rd [Formal opening, November 17, Robinson 1950.] Julie was about the eighth baby born in Oak Ridge but she was the first baby to be born in the Oak Ridge National Laboratory. Before this, already, the Whitakers had given us the baby carriage they had for their two girls, Margaret and Kathy. Kathy was a retarded child, the first one I'd known personally. They gave us a little apartment washing machine, 20 years old, that they had used in New York before they came to Chicago. Before this, this had belonged to Edgar Murphy. And he had offered to put a car at my disposal for anything involving legitimate family needs because there were no telephones in Oak Ridge; there was no bus service. The only communication we had where we lived, which was about a mile from the center of town, before the busses were installed (the busses came in probably about early January, 1944) was an Army jeep which circulated every street of the town every hour to take care of fires, unexpected births, and miscarriages, and various things which plague a town of 10,000 which was rapidly growing to 80,000. The Whitakers were just awfully nice in these regards.

They entertained us a few times at their house. Only once or twice did they come to anything at our house, but this was because, I'm sure, he worked a 90-hour week all the time. I remember one time when his health broke (I can't date this exactly) his wife forced him to go down to [Savannah] and Georgia for just a few days. That's the only vacation known to me he ever had. He travelled to Washington and Chicago an awful lot, but he travelled relatively little to the scientific conferences that happened once a month in Chicago, the Project council, because he was essentially involved in administration.

The science of his office was done by R. L. Doan, who is presently director of the Phillips Petroleum Laboratories at Arco, Idaho (and I've talked about Doan repeatedly). Doan and Whitaker lived adjacent to one another in "D" houses on what was called Snob Hill, and next to the Doans, lived Robert [Spencer] Stone, who was the medical director of the Clinton Laboratories. But there was another Snob Hill where the big shots of the Army lived.

Doan and Whitaker, who were utterly different personalities, must have had many conflicts. Nevertheless, Doan was a loyal supporter of Whitaker and worked on him privately in a friendly way. They travelled always, when they were both in town, back and forth to work in an Army car even before the 7 o'clock busses. They had the Army put at their disposal a Packard or a Buick. There was always a fight in the Army about what kind of car you had, and a laboratory director rated as much as a chicken colonel. [Military slang for full colonel from eagle insignia worn on shoulders.] There was no general in town, you see. They also returned late, about 6 o'clock. I'm sure they violated the rules about no secret papers at home. Of course, people could take secret papers on trips. Occasionally they'd see me trudging for the very late bus and I'd get a ride home, or once in a while they'd take me in, so I—I know in these discussions they often talked about general lab policy that wasn't ultra-administratively classified or Los Alamos-classified, and I could see that each one was totally informed on everything the other knew. A letter that went to either went into a community file that both read, and it was quite an effective arrangement. So if anybody wants to find out about Whitaker's war career, I think the person to go to is Richard L. Doan.

Right at the end of the war Whitaker was offered the presidency of Lehigh University, and most of us people felt it was a well-deserved position. I felt he lacked inspiration, but as I reflect now, and also remember the conversation Grace Mary and I had down at West Dennis [vacation spot, Cape Cod, Massachusetts], it's obvious that he had a position where inspiration would have been killing. He was not an Oppenheimer. Oppenheimer, you see, had enough inspiration. He could lead a laboratory and the Army recognized that Oppenheimer's spirit was indispensable. The Army in Oak Ridge didn't have enough sense to recognize that spirit is indispensable. The highest ranking Army officer in Oak Ridge was [Colonel] Kenneth Nichols, the coldest-blooded man I've ever met—a very intelligent man but completely without human feeling. I think that he and the crowd around him pressed on Whitaker this same cold, remote attitude.

As far as I know, Whitaker did an excellent job at Lehigh University. The *Times* spoke highly of him. Once, four or five years after he went to Lehigh, he was at MIT for some conference of college presidents, and he came up to my office and talked to me a few minutes, very warmly, and then phoned Grace Mary. She's always been very grateful for this.

Thinking of Whitaker reminds me of two typical rebellions at Oak Ridge that are not technical at all, and I just want to put these down because they represent the human element of Oak Ridge.

We lived in a tightly closed society. We were not part of Tennessee. There were thousands of Tennessee workers doing the steam fitting and the like, but they lived outside. People commuted from as far as 100 miles away. The various laboratory areas were strongly inter-compartmented. Men like Whitaker would know the administrative problems at Y-12, because he'd say, "You have to do it this way because Y-12 does it this way." We were the smallest project and the only project that was dominantly scientist and engineer. The personnel at the peak of the Clinton Laboratory, which I would put at about January 1945, was about 2,500. The personnel of Y-12, at peak, which I would put about July 1945, was about 25,000—mostly girls scraping uranium off of graphite, using code names and not having the slightest idea what was going on; or men, so-called physicists, who were watching dials to keep them in the narrow ranges and signaling for someone to come on a bike—no, bikes were K-25. K-25 had, I think, at its peak about 35,000, its peak was also the month of Alamogordo, Hiroshima, Nagasaki bombs. S-50, I suspect, was about three or four thousand, and not at all scientific.

Because of the high science components, and the high restlessness of scientists, and the high degree of intercommunication either between DuPont engineers and scientists (after all, we were aware of our differences, but our similarities to one another were much greater than the Army stereotype of what we should be and the typical personnel, supervisorial or lower, at the other labs), there were always fights. I've lived in two closed societies in my life. I've lived in Germany for a year under Hitler, and I've lived in Oak Ridge for three years under the Army. It's the same story.

The rumor level in a closed society is fantastically high and relatively accurate. Things go by word of mouth like mad. I think I've told of a number of cases—General Groves as Mr. X in the hospital [he was housed in a single room on the maternity ward for security], and Arthur Holly Compton as Mr. Holly. The other thing is that the methods of expressing resentment are extremely optimistic in such a closed society. There was resentment about the pattern of housing distribution, and this often worked to the great disadvantage of the Army. The Army, in the old days, set housing according to status in the industrial society they were designing. They equated a certain position as a colonel, as a captain, as a lieutenant, and so on. Then, a big wave of Army men would come in and there would be too many chicken colonels or majors with seniority, or something like this, and all the good houses would get sucked up. Then, there would be a storm in whatever lab was having most personnel move at that time, and through the personnel office and the housing office, the director of the lab would raise Cain. Then, the Army would pass an order out, "No more Army officers would get houses in Oak Ridge," and then, for three months no Army officer would get a house, until the Army got below balance. Then the Army would quietly let people in.

My friend, Lloyd [Robert] Zumwalt (1914-1998) came to Oak Ridge as a captain just when one of these stop orders went out, so he and his wife and their twelve- or thirteen-year-old son, Scotty, had to live in a lousy sort of a rooming house—one room—in Knoxville. There was no door key. He used a coat hanger. And they had to stay there about six weeks, until finally they were let into the poorest kind of separate house, an "A" house when the thing broke.

I think that probably half the time a man like Whitaker was fighting these personalities in these competitive problems among the labs. We were the first lab to come to full staffing, because of the urgency of getting the reactor going. Indeed, our lab began to have technical personnel in August 1943, and the reactor started in November 1943. So we were the first in Oak Ridge to produce our target, although our target was a gram of plutonium a day, which wasn't going to make any bomb. It was going to give Los Alamos plutonium to determine capture cross sections and fission cross sections and scatter cross sections and the various things that your father [Kenneth T. Bainbridge] was involved in.

We were riding high then. But the Army felt it had to watch expenses, and they pushed labor freezes. But, of course, more and more new things came up, such as our hot lab. Our hot lab was built after the deadline for all construction at the Clinton Laboratory. According to the elaborate plan, personnel would rise to 1,200 when the pile started. They would make material like mad, there would be a contingency for emergencies, and then they would eventually build down. Nothing was ever built down. In the labor shortage of 1944, I began to get the GIs, and I think I talked about Glendenin and Strickland and the crowd there, because there was a civilian manpower freeze. But the Army could transfer if we had a good squawk, soldier personnel.

Most of the plants, except X-10, used soldier personnel for the duller sort of tasks, the tasks that civilians didn't want. But at our place, in most of the groups, they were integrated according to their level of training, which was about average. Most of them were bachelors [degree-holders], or juniors with a science engineering orientation, with some Army special training in chemical engineering. Pretty soon, after they'd been around for a few months, since the game was so new, they stratified in the society according to their mental ability and their learning power. Some of them did extremely well. Then, we had to fight to get rewards for these men, and it was a hard job to compete for staff sergeants or master sergeants, or stuff like this that some of the men really deserved, in competition with the straight

Army who had an entirely different set of standards, where a regular Army captain felt it was his prerogative to create so many sergeants so often, and we had to fight the Army Personnel Office, which took into account the vast number of soldiers in Oak Ridge and not these one or two hundred rather selected men who had wonderful integration with this exciting program.

You always wanted to rebel at this stuff and fight the Army. Every time we had a chance, we'd fight the Army. But the Army never came in contact with us. The only Army man permanently in X-10 was Murphy ["Major E. J., of New York City, who was operations officer of the Clinton Laboratories and related research and experimental work," (Robinson, 1950, p. 158)], who was a really nice guy, plus three or four sergeants who worked with him. I suppose there was a security officer—the kind who hired the man who heard me talk about illinium-uranium on the bus. That wasn't done by Murphy, because Murphy would have checked with me before it went up. It came down to him from the top, from Major Perses [W. B. Parsons, Ibid., p. 157], who was the overall security officer for technical security in Oak Ridge. I think that Murphy's job was essentially promoting scientific work, following the Army's interest, and, more importantly, informing the Army what was going on and did this make sense and did we need the hot lab. I think I forgot to tell you that the hot lab came down that day, and it was only when the DuPont Company acceded to our having a hot lab could the science pressure for the hot lab be accommodated. I made certain small bargains with Greenewalt about how much plutonium analysis we'd do for him in exchange for having a hot lab, and all of this worked out fine.

Then the DuPont Company used the authorization for this hot lab to take care of three or four things. For instance, the drafting department had to have some new partitions and all, so this became a sort of a boondoggle. It was the last major construction going on in X-10, and the net result was that this building had a nominal cost of about half a million dollars. My guess is that the actual cost was about half this and the rest was just miscellany. This isn't cheating, it's just the way you get things done when you have a rigid Army-conducted society.

In the social society of the rigid Army, we'd rebel whenever we got a chance. One of the major rebellions, that cost Whitaker many a painful hour, was the policy in respect to busses. When we first came to Oak Ridge there had to be busses to take people from the dormitories and houses thirteen miles out to X-10. K-25 was further out; Y-12 was about four miles out. There were huge cattle cars used for the long runs, and messy Army busses used for the other runs. The busses were run according to the Army's idea of how people should work.

The main fleet of busses for our lab left town at about 7:10 a.m. It circled town to pick people up and went out. You'd get the same sort of bus back. If you missed the 7:10 bus, they made the next bus exceedingly late—9:30. Then there was a bus around noon, and around 2. There was a big fleet of busses to bring in the swing shift, which started work about 4:30 p.m., and then another fleet of busses around midnight for the graveyard shift. The X-10 didn't have much swing and graveyard shifts except for the plutonium reactor and the separations plant. The majority of people were day.

Did I tell you about signing in on the busses?

JBS: No.

Coryell: Well, the FBI once in a while would make a check of Oak Ridge, until somebody decided there was too much malingering in Oak Ridge. So they decided that if anybody went on an out-of-hours bus, he had to sign his name so that if a man's name came too often one could check with the supervisor was he doing his time.

This made us boil. I was one of the leaders who felt that the bus situation was an administrative horror and we ought to have flexible service that corresponded to business, and that it was up to the supervisors to see that men worked. And that if a guy worked till 8 or 9 p.m. at night and then had to wait another hour and a half for the 10:30 bus that he had a right to come in on the 9:30 or 11:00 o'clock bus, better than bringing him in on the 8 o'clock bus. The Army said this couldn't be done because it would affect the morale at Y-12 and K-25, where they had these thousands of men and women who didn't know what they were doing and were just wage slaves. Finally, I got on a bus at 9:30 one morning and the driver passes around a sheet to sign your name. I said, "What is this for?" and he said, "You have to sign your name. It's Army orders." So I calmly marked down, "I refuse to sign my name. Charles D. Coryell." Two days later they dropped it.

This was also involved in the fight about the same time about time cards. The Army insisted, and their scientist leaders, men like Whitaker and Arthur Holly Compton, had a fair amount of in and out of Oak Ridge and had a fair

344

amount to do, I think, with liberal decisions made. He had a charm, and he could handle Army officers when Whitaker's quiet, hard plodding met stone walls. Finally, the Army forced the issue that the scientists would have to sign time cards.

There was a major rebellion. There was a mass meeting. We were told we couldn't hold a mass meeting within hours, so we held the mass meeting out of hours. This was the worst sort of insult you could give to a scientist. The DuPont engineers also felt this as an insult because DuPont has lots of time cards for the workers but anybody who has a supervisory status is in the category he runs his own time, although the DuPont people are always 8 to 5 and always had clean desks, and we worked the most fantastic hours and had the messiest desks that the Army and the DuPont people had ever seen.

Finally the decision was made that each supervisor would initial the time cards for all his men, I had about 40, and would mark 8 hours for each of the 6 days of the week, whether the man had worked the 8 hours or not. All we were doing was signing the man was doing a good job and that on the average he was working at least 48 hours a week.

This made me angry, also, because I had a period when Louis Stang worked 6 weeks in a row, more than 80 hours, and I put down his hours. Anybody that worked more than the usual week, I put this down as an extra thing. This also would make the clerks mad, and I'd get phone calls. "What did you put this down for?" And I'd say, "By God, he worked that way. You'd better leave it that way." It probably got changed to 48 hours in a higher office when they put the sheets away for the Army.

There was an absentee contest a little before this, which was called a presentee contest, and the Clinton Laboratories won, naturally, because we certainly had the best morale in the place. The science crowd all knew each other, and the mechanicians, and the accounting room girls and all. On the whole, they felt integrated to their groups and they had a lot of loyalty to their groups. So Clinton Labs won hands down. We had a very low percentage of absenteeism compared to the big plants, where there was no spirit and terribly dull, obscure, and fantastically unreasonable jobs. The girls there were never told why things were done. They were just told, "You do it this way." It was totally arbitrary.

As part of my campaign of harassment, I wore my badge upside down, systematically, as a protest against the bus service. In spite of this I worked more hours than the other people. I took pains to see Whitaker and Doan, and ask, "Why don't you ask me why my badge is upside down?"

The bus rebellion was one way I could express my feelings. There were busses put into town, and these were also free. Finally the Army decided you can't have a town with free bus service for everybody. This is contrary to the American method, and Congress will raise hell. So the Army decided arbitrarily, probably in mid-February 1944, that there would be a nickel charge on all bus rides.

A storm broke in all plants. Everybody who spoke on the thing said, "Our contract says we have free transportation to work." So the Army finally compromised on a nickel a bus ride in town. They had to print little paper tokens because metal was impossible to get. I have a few of these as souvenirs—lovely little things. If I find mine I'll put it in the file. We had a mass meeting in the village about this, but the compromise was no pay for going to work because many contracts said that. I don't remember seeing my contract. We were University of Chicago boys and this project was a continuation of the contract we had in Chicago and busses weren't mentioned, but this wasn't true of people recruited for Y-12 and K-25. They probably had contracts which specified benefits—hospital and things of this type.

JBS: But you did have to pay if you were just riding in town rather than out to your—

Coryell: Well, after—Yes, the Army compromised that housewives and people doing business shopping in the village would pay like a normal town. You see, Oak Ridge was about seven miles long and about three miles wide and lots of people didn't have cars. There was pretty generous gas rationing for us. Anybody who had to commute from the outside had it, but all of us—I had a car from 1943 on, a 1936 Chevy, a terrible old wreck, sounded like a tank, and I never had a shortage of gas because I never had time to go on long trips, but some of the people who went to Gatlinburg every weekend, and some of the wives had [blank] people and just had to get out to some place away from the pressure of Oak Ridge— they could borrow or get gas rationing from people like me who were more heavily occupied. I loaned my car several times to young people to take a whole load of people off to Cumberland State Park or things like this I never did see.

JBS: Weren't there restrictions on moving out to cabins in the mountains and so on, in the way that there were at—

Coryell: No. Los Alamos had this sort of business that scientists couldn't travel without another scientist with him, like two nuns. We had no restrictions of this type. We were in a closed town, we were in a dry county, the Army reservation was dry and there was no officer's club and of course everybody was anxious to bring in liquor. But on the whole, I think one suitcase full of liquor in 50 or 100 got picked up. But there were no restrictions on our moving out down South, though were I to go on a scientific trip to Ames—any trip involving a university laboratory or a laboratory associated with the Manhattan Project, or any trip involving a university that they happened to find out about—I would have to talk to a security man before going. And any trip to a classified lab, there had to be a B-2 paper. About half the time these papers would be fouled up.

Oak Ridge was a place where the standard Southern administration help was grossly incompetent. Fifty percent of the time prominent visitors would have to wait at the gate while the thing was handled by telephone. Half the furniture moved to Oak Ridge came the wrong day and went to the wrong house. All this helped drive the women wild.

We had a house promised us in September 1943, a lovely place at 137 Outer Drive, near the corner of Michigan, a "B" house, second from the bottom, the nicer of the two-bedroom houses. This had a lovely porch over the trees. All the houses entered on the side away from the street. The back door was on the street. Have I talked about this house?

Coryell Interviewed by J. B. Safford, 4 September 1960

JBS: Not about the "B" house. Didn't you move into an "A" house?

Coryell: No. What happened is that giving this tape recording has stimulated my memory, and Grace Mary and I have been talking about a lot of these things. Julie is quite interested, so we're telling her. [Julie was home from a month-long tour in France and Belgium to practice French. She had met Jeanne and Moïse Haïssinsky, at their home in Sceau, suburb of Paris.]

I was down in Oak Ridge for a quick look-see about August 23rd, 1943, and trudged up to see my house before there were busses in. It was nearly done. James Franck had already visited Oak Ridge earlier and he, too, just to be nice to Grace Mary, had walked over and told her it was a lovely place. People were worried because there were rumors in Chicago about the awful flats. There were some very dull places, but on the whole they did a beautiful job. We had a house in the first thousand. They were nicely architected and they've been described in *The New Yorker* by Daniel Lang as early alphabetical [blank]. His book, *From Hiroshima to the Moon*, I think, has this saying in it. [*Cooking Behind the Fence* (2003) illustrates modular homes A – D with three bedrooms, plus F for flat top.]

I came down September 23rd with Henri Levy and planned to bring my wife October 1st, when our lease expired in Chicago, and our house had not had a thing more done on it. An emergency somewhere else in Oak Ridge had taken the construction people away, and it turns out that the contracts for the second, third, and fourth thousand houses were all

349

done. The first thousand had bogged down in some administrative red tape in Washington.

Did I mention how the furnaces were illy [*sic*] designed in the houses?

JBS: This was a problem at Los Alamos also.

Coryell: I'll tell you that a minute later.

I had made a pact with Harrison Davies, who was in my group, and David Hume, who was in my group at Chicago. Davies came in about March of 1943, and Hume came about June. Whoever of us got a house first would put the other two couples up. Harrison Davies lost this pool. He got a "B" house on New York Avenue in the sun, but with a lovely view on a hill. The Humes had a very pretty house right back of the chapel down near the center which was finished last. So the Davies had the Coryells with them about four weeks, and the Humes about six weeks. The Hume house, when finished, proved to have hornets in it and Aloyse [sp?] Hume wouldn't move in it until the hornets were proved to be absent by another week or so after work. I guess the Humes got the hornets out by themselves by putting chemicals in, rather than by counting on the Oak Ridge maintenance service.

Our house was finally ready to move in by about December 1st. I did a lot of finagling. I pulled my rank often on DuPont personnel. That's the only time in my life I ever pulled rank. I insisted on being informed when my furniture was arriving because I had a pregnant wife and we had to get settled. They said we could have possession of the house but that there was no moving truck available for about two weeks. I fought and fought and fought, and finally checked with Chicago and found that was right. There was no more chance of our getting furniture from Chicago to Oak Ridge, for about two weeks. So Grace Mary and I, and Henri Levy, and George Packer and about four or five others did a complete job of waxing the floors. We darkened the floors very much. By the way, these floors were hardwood. The houses were very lightly built but very attractively designed. They've been taken care of very well and they still look very good.

The following week, on a Sunday, we had another party. Someone got beer for the kids, and we did some more waxing. The next day I got a phone call from this guy, Rose, who'd sworn the furniture wasn't coming for another week yet. "Your furniture is in Oak Ridge." Meanwhile an Army officer had located Grace Mary at the Davies and taken her over there, and she was furious that the DuPont's subsidiary was moving in all this furniture on these sticky floors. Grace Mary hated the DuPont Company as much as she hated the Army from then on.

Whenever you had anything wrong with your house you'd finally find a phone (there was a phone about every mile on the streets) or you'd pass a note by this jeep that came by, and six to twelve people would come to take care of it. I never saw such manpower put on such stuff. We had trouble with our furnace. Julie was born [Friday] December 24th, and I had to go off to Ames in early January. While I was gone our thermostat failed. It was a terribly cold period and Grace Mary had to get up and shovel coal every few hours. Coal was furnished free, and it was awfully low-grade coal. There were coal bins, and always they'd throw in coal on top of your shovel and things, which would get buried down there, and you'd have to go and borrow your neighbor's shovel to dig it out. You'd phone if you were in trouble. You were asked not to call for coal until you really needed it because they hoped to take care of it so that if the weather changed you wouldn't have to worry.

Grace Mary got her first contact with illiteracy. She phoned down around eleven o'clock one night, and the man said, "Sorry, lady, I can't do a thing for you. I got to wait until the morning man comes. I can't write."

She tried repeatedly to get this thermostat fixed, for about ten days. And finally I went next door, where a house was temporarily vacant, and stole one. Things worked fine. When the people moved in next-door we told them we'd stolen their thermostat, so they went down the street and stole one out of the next place. Three weeks later six men came to fix the thermostat, which had broken two months earlier. Grace Mary told them we'd stolen one, and they said, "You should have done that in the first place." They didn't worry about going down and finding the house that this must have bounced to.

There were several serious fires. The furnace was a coal furnace, with an air blower and a fan. In summer you had cooling. Of course, the fans blew all sorts of coal dust through the house. The furnace had about a half-inch clearance from the ceiling of the house, and if somebody threw in too much coal, wasn't used to using coal (and I was one of these people), you'd have a really hot fire and you'd set the house on fire. If you set the house on fire, you'd try to put it out and get help, but help came pretty fast when there was a fire. But the contractors continued to put in these furnaces. It was cheaper to put them all in, than to get a work order through Washington to change all of them, than to change the specifications of the houses that were being built. Just typical of the wartime operations.

JBS: I wondered if you would say something more about Nichols, who has become famous since his District days.

Coryell: I met Colonel Nichols several times. I don't remember his ever giving a public speech, though Groves sometimes gave public speeches—generally insulting the people he was speaking to. At the end of the war all these soldiers got an award for merit, a golden cluster, a wreath sewed on their uniforms. I've forgotten it exact name, but it was called by the soldiers the "golden toilet seat." It was just the shape of a toilet seat, and the boys claimed they got it for lower than average venereal disease.

The day this was granted was about three days after the Japanese surrender, and General Groves came and gave a speech. He gave moderately warm praise to the scientists and engineers, and their soldier helpers, and all. Then he said, "Except you, you, and you, and the other sons-of-bitches among you." He always ended any praise with a nasty twist of the dagger, which made people mad and they worked harder. I think this was a calculated thing.

Did I mention Groves running? He'd come in on the 5 a.m. train and run till midnight, with all the colonels and captains and everything else running around behind him. He'd storm into the office around 7:30 or 8, and the day Groves came all the Army officers came and worked before their 8 o'clock time. He comes down through the hall, sees Lloyd Zumwalt working at his desk, says, "That captain needs a haircut," goes down another line.

Nichols was the administrative man at Oak Ridge, and I considered him a man of very high competence. As I mentioned in my senator-science thing, he was the only Army officer I knew with an earned PhD in hydrodynamics. But he was colder than any other man I've known. Whitaker had a coldness which I now interpret differently than I did at the time. A[rthur] C[lay] Cope, at MIT, has a coldness which is lots warmer than Nichols.

I went to see Nichols right after the end of the war to ask his advice about the technical situation. Could the Y-12 plant, or the K-25 plant shut down, because the atomic scientist groups were urging the United States to offer a 6-months' moratorium on production of atomic supplies coincidental with the Baruch proposals to the United Nations on June 10, 1946? This must have been about May of 1946. Nichols was nice to me, but extremely cold, and just said, "Technically impossible. The factories would deteriorate so fast they'd be worth nothing." I'm sure this is not true, but I'm sure this is the way he thought and it was his story. He was civil, but he got rid of me without transferring any of the type of information I would have liked to have had, a sense of balance in the production side, because I knew the plutonium project extremely well, but I had no knowledge about the electromagnetic process and the gaseous diffusion. Electromagnetic had already been shut down, so we knew this must have been on economic grounds.

I met Nichols that night where I embarrassed him by asking him to speak in the presence of a superior officer, and he just blushed and sat down.

As long as you brought up why you wanted to speak about Nichols, suppose we go into the Oppenheimer case. You've been busting to get into this.

JBS: O.K.

Coryell: I talked last time or the time before about my contacts with Oppenheimer in Washington, and how I felt that I understood why he supported the May-Johnson bill. I had enough contact with him to just come to worship him.

I saw him very rarely after that, occasionally at the American Physical Society, but I remember a cute statement of his: "You can judge the importance of a man by his diffusion rate at a scientific meeting." That had just got into *The New Yorker*, I think. Somebody clocked how long it took Oppenheimer to get across a foyer at a Chicago hotel where the American Physical Society was held in 1951, and computed in feet per hour the migration time. Matter of fact, everybody wanted to see Oppie. He was very well informed and very helpful to almost everyone I knew, but I knew several people who felt that Oppenheimer was an arrogant man and one of my very good friends thought that Oppenheimer had done dirt to a couple of people. I argued with these particular people, I said, "How do you expect a guy to go on without making mistakes?"

I inquired a little more about the case where Oppenheimer had done dirt, and it seemed to me Oppenheimer had made a hasty decision that this man was no good as a physicist or something like this. It seems to me that a man who has to evaluate other people has to act on prejudices. Either you say everybody is perfect, which isn't true, or you say some things which may be too far the other way.

I was satisfied that Oppenheimer had the interest of the country at heart, and also I remember that in the mid-forties—1945, '46, '47, '48—Oppenheimer was against ideas of the type such as stopping production for six months. I don't remember his reaction to this specific one, but I think that had he been for it in any strong way there'd be plenty about it on the record. Oppenheimer's position, in every contact known to me personally with him and through other people, struck me as a very conservative position with respect to military preparedness and a very liberal position with respect to somehow or other getting negotiations that were meaningful. Therefore, when I first heard about Oppenheimer being charged with being a Russian spy, I was horrified. I think that had there been any falseness about his support of the United States production and things like this, it would have shown in some of the many contacts that we had in the Federation of American Scientists with the thing. Oppenheimer was a god of the Federation of American Scientists, and they supported him, I guess, more than any other organization during the Oppenheimer fight.

I was in Israel as [a Louis Lipsky] Fellow [at the Weizmann Institute of Science, Rehovot]—I think all this stuff is in the record here—from December 1953 to June 1954. In April, when the Oppenheimer case broke, two separate friends of mine—Lionel Goldring and Julius Bregman [?] (Lionel being at Brookhaven and Juli being post-doctoral at MIT)—sent me

the pages of the *New York Times* with the indictment of
Nichols. I was horrified when I read this. There was the cold,
hard Army officer bringing every imputed evil or error in
Oppenheimer in 22 or 23 hard, terse paragraphs, with not a
single statement about the singular contributions of
Oppenheimer. However, the Atomic Energy Commission,
since 1948, had as public policy to judge a man by his whole
personality and his whole record, not by any given part of it.
I don't think that the citation for the Rosenbergs, which read
their death sentence, was probably any harsher than the
citation for Oppenheimer.

I resented this very much in Nichols and Strauss. I think
Strauss, also, is a hard man. My knowledge of Strauss is
almost all secondhand. I met him but twice, and had a
friendly correspondence with him in 1948 when he opposed
the sending of radioisotopes abroad because a reported
Communist who worked in the Norwegian navy or army
was one of the recipients of iron-59. Most of the scientists
were so angry at Strauss they wrote hot, snotty letters to him
and I wrote a letter trying to point out the good and bad in
things like this, and I got from Strauss a long, intelligent
defense of his position and a warm invitation to come to see
him. He was then just a member of the Atomic Energy
Commission, a Truman appointee. I responded to this and
added some more information in the same general
framework because the Hickenlooper search for lost
uranium, I think, was going on about the same time, and I
got a similar warm answer from Strauss. I always figured if I
did get to Washington I will call this guy, I'd like to see how
he ticks and I think I can differ with conservatives effectively
and I learn by it, and I think they do some.

I didn't know much more about Strauss' crookedness and arrogance until I saw it develop as I saw the Oppenheimer case. My sources of information in Israel about the Oppenheimer case were the *Jerusalem Post*, an English language daily published in Jerusalem then by Gershon Agron (1894-1959), the uncle [brother] of Martin Agronsky (1915-1999), and one of the best newspapers in the whole world for effective news transfer and an effective Adlai Stevenson political position, except it was what Stevenson would have had as a general scientist in Israel. The international position was perfectly Stevensonian. It was a very effective way to get world news. The quantity of information they published on the Oppenheimer case was rather small, because there was little news, except that interest in Israel was high. But I think that people in Israel worried whenever a Jew was involved in a subversive charge because they instinctively worried that anyone should damage the position of the Jewish race in the world, and they know that there are Jewish sons-of-bitches. Secondly, they consider it probably partly an expression of anti-Semitism and Jews are very cautious about speaking of anti-Semitism unless they can take a very safe position, because it just feeds the fires of anti-Semitism. So, my present recollection is that the *Jerusalem Post* would carry, until the Gray Committee report came out, probably not more than 100 inches in a series of small articles.

The Gray Committee report was published by Strauss on a technicality. Strauss reported that [Commissioner] Eugene Zuckert, a man I've since come to know rather well—and I'll inquire about this matter the week after next—Strauss said that Zuckert lost it on the Washington-New York train and that for three hours his briefcase was missing with the

Oppenheimer testimony in it. Strauss therefore decided that this probably had been seen by a newspaperman, and that in the interest of fairness he'd had to publish the whole thing and he hastily declassified and published it. This was nothing but a crooked ruse.

JBS: It's true the thing was left on the train, because Harry Smyth found it in the Lost and Found office of the station.

Coryell: Sure, but the presumption that a newspaperman had seen it when it was only gone a few hours is nothing but a phony assumption. Had I been Zuckert I would have raised hell because Strauss is such an arrogant man that he thinks that if he and I know the same facts I will think exactly as he thinks. He has never had the breadth to see that people can look at the same facts on both sides, like the Republicans in the room and me, and come to utterly opposite conclusions. He's one of those daddy-knows-best people who thinks that when daddy tells you to go, OK. Strauss damaged his position greatly by this Oppenheimer hearing.

I don't remember the Israeli response to the Gray Committee. This must have happened in late June, and I left Israel on June 27th. I was just shocked to death by the two copies of *The New York Times* I got. I was in Haifa, at the meeting either of the Physical Society or the Chemical Society. Both met in the month of April, one of them over the Passover weekend, and I immediately wrote a warm letter to Oppenheimer, telling him how sorry I was for all this and wishing I could do something to help him.

About two weeks later I got an airmail letter from Lionel Goldring, telling me he had his security taken away from him. I'm not sure I knew then, but I know now, his security was Defense Department, not AEC. His work, I think, was development of [blank]. He had named me as the man who would be a character defense of his and who knew more about the charges against him, which he outlined to me, than anyone else and would therefore be an important member of his case. Would I send an affidavit? And would I also agree that shortly after my return to the United States, I would attend his hearing? This I did. I sent a telegram right away, and my next job with him was to get a notarized statement in Israel. It took me most of Thursday evening to write a three or four-page detailed statement of my knowledge of Lionel Goldring and to address the facts as I knew them from his letter about what the charges were. I got this finished, typing myself, Friday morning. Israel closes up Friday noon for the Sabbath, and about 10 o'clock I borrowed the lab car and driver to go down [to town] to try to find a notary public. (It turned out that the lab didn't have a notary public.)

I assumed that Israel has three kinds of law: Talmudic law, Ottoman law, and Mandate law—which is British. I assumed there would be enough residuum of British law that the notary system would work. I also knew I could get this notarized at Tel Aviv at the American Consulate, but it was trouble for me to get a car to go there, and there was danger of their closing at noon and it was impossible to make a telephone call in Israel that far away without waiting a couple of hours. So I went to a Judge Moscowitz, who was trying an Arab for theft, and he stopped the case and called me into his chambers and talked the case over. He was very suspicious because if he gave proof that I swore to what was in here, and this was going overseas, here I was invading the rights of the American Consulate or the Israeli Foreign Office, and he would have to refer this to the Foreign Office in Jerusalem, which couldn't be done before Monday. I persuaded him that all I wanted from him was a statement that this was something I swore to be true for use on behalf of a friend, and that my swearing is a public statement in the United States and this would be the equivalent to a notarization. So he agreed that, by some little formula—he was quite competent in English—and then he called me on the docket where the Arab had just been removed to wait until the judge came back, and asked me whether I preferred to swear on the Torah, the Bible, or the Koran. I selected the Bible and swore that the statement was true, the whole truth, and nothing but the truth, and sent it off to Lionel. I barely got to the post office to get the stamps to take care of it. This was several Israeli pounds of postage.

About two weeks later I received a similar letter, but not so detailed, from Walter Rosenbluth of MIT. He, too, had security problems. His charges were identical with Lionel Goldring's. They involved the Association of Scientific Workers in Cambridge, but in addition they stated that a meeting at his house, on a certain night in September 1951, had been in fact a secret meeting of the Communist Party.

Now it happens that I attended that night. This was an informal meeting of the so-called executive committee. There were about 12 or 14 people there, and the guest speakers were two members of the Association of Scientific Workers—Harlow Shapley and [Kirtley Fletcher] Mather (1888-1978). They were specifically reporting on the [Dirk

Jan] Struik (1894-2000) case, both had visited Struik in jail. This dates to within ten days after Struik's arrest. A new member of our staff, a new assistant professor in physical chemistry, David [P.] Shoemaker (1920-1995), had just come from Caltech. I knew he'd been active at Caltech in the Association of Pasadena Scientists, the FAS branch in that area, in his day, and I asked if he'd like to come. I told him he'd meet a bunch of nice people and it was sort of interesting, that the case was a big issue in Cambridge and I was fairly sure that the MIT position was strong. So he agreed to come, he didn't have anything better to do. I didn't bother to tell anybody about his coming. I brought him. I said to Julian Waldo, who was with us, "This is one of my friends, David Shoemaker, whom I'd like to have come along." There was no objection.

It happens that this act, this accidental act of mine of bringing an arbitrary person, unknown to anyone else in the room, was the main evidence that this thing was not a meeting of the Communist Party. For this reason I was an important member and so here again, the hearings would be postponed until I came and I would be asked to take a prominent position in the hearing.

I then deduced that everybody who was liberal and outspoken and, in particular, everybody known to be friendly to Oppenheimer, was in trouble. I thought this was a security shambles, and I thought, when I go home I'm going to be defending all these guys but I, too, am going to be having hearings. They're waiting until I come home because I have a sort of leave of absence. My trip had been authorized by the AEC and I agreed to report back to them anything odd I'd seen and I was required, on entering any new country, to report as soon as possible to the highest ranking United States official I could find. This got me into some other messes I may come back to in a minute. But I was extremely security conscious.

Then we read the Paris edition of *Time*, which came the day after it came out in Paris, and we got *Life* and the *New Republic* from the United States about a month late. These were my sources of information on the Oppenheimer case and, of course, there was relatively little. But in the middle of the Oppenheimer case, *Time*, by leakage sources, got enough information to decide that Oppenheimer was evil. He was one of their great heroes, *Time* and *Life*, and he suddenly turned from being a great hero to just a vile man. The transformation in *Time*'s tone and all was just awful. I resolved as soon as I got home I was going to cancel my subscription to *Time* with a hot letter, which I did. I never got a rebate on the remaining month or two they owed me, either. This was [Clay] Blair, [Jr.] (1925-) and [James R.] Shepley (1917-1988) at work. Shepley is presently an important advisor to Nixon and is involved in this African case. I wrote this to Kennedy the other day, that this man is the man (Shepley and Blair) who is responsible for this evil book called *The Hydrogen Bomb*: [*The Men, the Menace, the Mechanism*, 1954]. This book came out in 195[4], and it gives all the "facts" in the Oppenheimer case and ends with the statement of the authors that they're sure that Oppenheimer is a spy.

JBS: You may be interested to know that this book is attacked on every tape of every person I've interviewed.

Coryell: I hated to give those bastards my money to buy their book, but I wanted to read it in a hurry. Then I marked in the front, "This is an evil book and anti-MIT, and I encourage everyone who reads it to read it with his eyes open." This stands in my copy of the book, and it was in my office in the general interest file for many years.

As I deduced from what I read during the summer, travelling—we left Jewish Jerusalem for Arab Jerusalem by the Mandelbaum Gate and I had a second visit in Jordan with my family, passed through Lebanon, and went on to Istanbul for a few days where I was a guest of the Technical University. Then we were at Athens a few days, over the Fourth of July, and in Rome about five days. Then I went to London and got a [Ford Consul sedan car] we still have. Grace Mary, Julie, Miles Sherrill and I travelled all over Europe. He joined us about the 17th of May, 1954.

I was under obligation at every country I entered to go to the United States official, and I decided this was not a farce. I had already told, back in Spain—I think I had better drop back to my own security problem because I identified myself with the Oppenheimer case and it turns out I was wrong. But it also made me acutely aware of many aspects of the Oppenheimer case that are perhaps worth recording.

When I planned my trip I had to fill out form 290, where you had to list everybody you intended to see. This is an impossible thing to do for a trip of ten months, but I listed the prominent scientists I knew and would certainly go to see, and on this list was Moïse Haïssinsky, a moderately left-wing French radiochemist who worked with Marie Curie a short while and who was one of the chemists with Fermi in

Rome, a Russian Jew who'd been a year in Israel in the twenties on a kibbutz and then had got his PhD in Italy. He's a wonderfully provocative and interesting person. I love him very much. Another man on the list was Pierre Süe, now dead, who was editor of a magazine called *Atomes*, a rather good radiochemist and the chief chemist of Frédérick Joliot. I think by wisdom I didn't put down Frédérick Joliot and Mme Irène Joliot-Curie, though I'm sure I met them in 1948 and I have a lot of respect for them scientifically and a lot of affection for both of them. It turned out I didn't see them because I was in Paris in the summer and I didn't try, because of the Oppenheimer situation, till I knew my ground. Another man often in trouble with visas, whom I should have put down, was [Michel] Magat, of the Laboratoire de Chimie Physique des Rayonnements, University of Paris. Magat's another problem I know about that's caused lots of State Department trouble. I think it's a crime the way the United States acts about him. Haïssinsky had been denied a visa several times by the United States in the messy days of the late forties.

Well, the AEC had sent a young man named [George F.] Quinn, whom I've since come to know rather well and have a lot of respect for, and his job was to persuade me not to see Haïssinsky and Süe. We had a long argument in Irvine's office where we first started reporting, with the door shut, and I said I would do anything the AEC said if they put it in writing, but that if I went to Paris and didn't see a dear friend at whose house I'd stayed in 1948—Haïssinsky—I would make more embarrassment for myself and the State Department because I couldn't leave this unexplained. Therefore I insisted on seeing them unless I was specifically prohibited. The next day a phone call came from the security officer in New York, a man not known to me personally, who said I could do anything I wanted to if I was very careful, particularly on dark nights and on empty streets. That was this cloak-and-dagger.

I was tremendously euphoric about this wonderful trip coming, and Grace Mary and I went with Marguerite Yost, the wife of Don Yost of Caltech, down to Washington to visit Fred and Julie Stitt (these are pre-war friends of ours, our daughter's named for Julie). We wanted to show Julie Washington, D. C., before she saw the famous capitals of Europe and Asia. This was an area in my life, as now, when science and politics tremendously interweave, though there wasn't any politics crisis I can remember. Then we set off on the [American] SS *Independence*. We left New York on the day of the Weizmann Institute banquet [December 3, 1953], an annual banquet held in honor of Chaim Weizmann, who had just recently died. This was the second banquet held after his death, a fund-raising banquet where $300,000 was raised. The speaker that night was Harry Truman. I was very anxious to hear on our ship's radio what Truman said, and I was disgusted to find out that on the high seas nobody ever pays attention to land affairs. You get a mimeographed letter. I never did find out what Truman said.

December 8th was the day that [Dwight David] Eisenhower (1890-1969) was giving a big UN speech on the peaceful uses of atomic energy, and I never found out what he said until about February in Tel Aviv. Then, as we were approaching the shores of Europe, we passed the Azores. I got worried we were landing in Gibraltar, our ship is late because of a strike in New York that made us late starting, we're landing in Gibraltar about 9 at night and we're due to leave for Spain at noon the next day, and I'm technically required to specify every country I enter, and Gibraltar, by my definition, is a country. So I sent a telegram at my expense to the American Consul in Gibraltar that I would be arriving on the SS *Independence* and would like to have a chance to see him first thing in the morning of the 11ᵗʰ of December. I got a telegram back the next day from the American Embassy in Lisbon that the consulate in Gibraltar had been closed, and that I either contact them or Madrid.

We were scheduled to be a week in Spain and were not going to go to Madrid, and so we went to Gibraltar. After all, I found out there was a Naval mission to Gibraltar, and the leading official of this mission is a Commander Wilson. I heard about him at the Rock Hotel, a very nice English Hotel in Gibraltar, so I looked him up at his home about 9 the next morning because we were trying to leave for Spain at noon. He was a wonderfully nice guy and he must have thought I was a real crazy scientist. He wanted to know why I had to do this, so I told him, "I'm required." We talked a bit and he turned out to be a friendly liberal. I told him what he should do is make a little report on me, which I wrote out, and get this typed up when his office opened—offices don't start very early in Gibraltar—and send a carbon copy to this guy [C. Arthur Rolander, Jr., *Biographic Register*, 1953], the [AEC] security officer in Washington (a man I still have never met), and a copy to Jeffries Wyman (1901-1995) [Ibid., 1954, from A. Twiss-Brooks], who was the science attaché at the Paris Embassy. (Joseph [B.] Koepfli was at that time science advisor to the Secretary of State. By the way, Walter G[ordon] Whitman (1895-1974) has just been appointed to this position now [1960]. He'll be the best man who's ever held it. I'm just thrilled to death.) But I knew Jeffries Wyman knew and liked me and would understand what was going on. I knew the security officer in Washington, no matter what I'd do, might hold this against me at some future time. But I decided this was a serious situation. In the discussion with Wilson it occurred to me that the reason the AEC wanted to know where I am is that if Senator McCarthy gets roaring in the United States Senate, "Here's red Coryell in Israel, or here's red Coryell running around loose in Europe, how will you find him?" The AEC would always know how to find me. I think this is probably fundamental, not necessarily McCarthy era but just in case there's a missing scientist, like the present mathematicians who are thought to be in Russia, they'll know where I am. I, religiously, for the rest of my trip followed this thing.

We left Gibraltar about 2 p.m. the day in question with a hired car—a man named Antonio Wilkie—a nice Gibraltar man, Spanish-speaking with a little English, who'd been at the British Embassy in Madrid during the war and was terribly anti-Spanish. He became a good friend of ours. I decided we would spend the first night in Malaga and I would check the American Consulate there. Malaga is a city of about 400,000 where oxen still go down the main street and there are barefoot people in front of the fancy palaces. So next morning about 9 o'clock I look up the American Consulate and phone, and a Spaniard answers and says the consulate's been closed and the nearest consulate is Seville. I didn't want to talk to a Spaniard. I didn't trust Franco Spain and I have many more reasons not to trust it since, including their stealing some films of ours. So we went to Granada, where there never had been a consul and after a few days in Granada, staying at the Parador San Francisco [recently converted from a nunnery, no electricity except in public rooms], a lovely time, we went on to Seville and stayed at the Hotel Inglaterra. I had already decided this was nothing you did over the telephone, this was a subtle sort of thing, so I went down to the consulate—we had a little car trouble. Grace Mary has never forgiven me because we missed seeing the Cathedral of Seville, where one set of the bones of Columbus is. The other set is in Genoa. The cathedral was closed from 12 to 2 and I spent the time on this goddam [T. H.] Chylinsky, [Ibid., 1951, p. 37].

I stepped into the consulate, the first one I'd been able to attend. Counting the United States and the United Nations this was my fourth country, but I didn't have to report in the United States and my visit to the United Nations was too short to report. I was with an American. The office was full of Spaniards and I had taken with me only one card, my Los Alamos card, which I only used for purposes like this. This card is a dramatic red card; I suppose you've seen your father's. I used it to cash a check on the Cape the other day, because it's not so secret any more. All the other stuff I left behind, but I had to have something to identify why I went in.

I asked to see the Consul, and said that I was asked by the Atomic Energy Commission to do it, and I was brought into a young, gray-haired man. Grace Mary and Julie, I think, were out in the office, and he was nice to them. But he was very cold to me and wanted to know why I had to do this. "Well," I said, "the AEC says I have to." He said, "Why?" Then I told him the story. I said, "This is in case Senator McCarthy asks where red Coryell is, they'll know." And he gave me an awfully cold reception, but he agreed to pencil out a little letter to go to Rolander with a copy to Wyman, and I went on my way.

In Egypt I was very nicely received, but there I went to the United States Information Service and I had an awfully good time. I told the man I was talking to (I don't remember his name right now) that I was speaking at Cairo University and I would like very much to have a United States government official hear my speech because I was talking on President Eisenhower's Atoms for Peace and I was going to speak as if I had read the speech but all I had was a two-inch item in the Gibraltar newspaper. They didn't have a copy in the USIS (this was only about five days after the speech) but this guy was awfully nice and I think he did send some observer to my speech. But you know, I figured there might be trouble and someone might want to chop my head off, as later on people did chop Oppenheimer's head off. At any rate, I liked the reception in Egypt. He told me an awful lot about the USIS and the troubles they had with McCarthy.

Then I went to Jordan and I had to see the American Consul. I was crossing war lines at the Mandelbaum Gate, Jerusalem. There's an American consul in Jordan right at the [Gate]. I was uneasy about going across a mined area. There was a real state of war between Israel and Jordan. We arrived in Jordan about December 23rd, at about 2 p.m., and darned if I didn't drop in the airport the only Hebrew alphabet I had with me. A Jordanian picked it up and handed it to me with a smile. [Charles had provided separate papers called *laissez-passer* so the Israeli markings would not be recorded in our passports, and we could return to Jordan at the end of our stay.]

We left all our books in Hebrew and all our stuff that would mark us bound for Israel in suitcases in Naples, and we'd persuaded two Israeli pilgrims to take them to Israel for us and we'd pick them up later. We were awfully tired. It was bitter cold and there was a little snow on the ground. We were put up in The City Hotel, a nice Arab hotel about half a mile from the Mandelbaum Gate. That night we heard explosions, and Grace Mary thought, "[Oh, oh], we're going to have war at Christmas."

The next morning I went around to the American Consulate as soon as I thought they'd be opened, but I thought I'd be a little easier than I was with Mr. Wilson at Gibraltar, so I showed up about 9. I was also very warmly received there. No, I'd gone down there that afternoon and the consul asked me what I wanted to do. I said, "Let me type a letter. I'll get a spare typewriter, and if you'll just transmit one copy to Rolander and one copy to Jeffries Wyman, it's all right." So I marked down "U. S. Consulate for Jerusalem in Jordan." It is a common way of identifying. You say Jerusalem in Israel for the Israeli side of Jerusalem, and you say Jerusalem in Jordan.

The consul reprimanded me in a friendly way. He said, "There is no U. S. Consul. There is no Jerusalem in Jordan in the State Department. The State Department accepts only the decision on partition of the United Nations of November 1947, which says that Jerusalem shall be an international city. It happens because of the war situation that we have to have a consulate on each side of the war lines, but there is really only one United States Consulate in Jerusalem."

I think this was the afternoon we arrived. The next morning I went back to see what the explosions were about and they said, "Never mind. They're just cleaning the mine fields, because about 10,000 Christian pilgrims were coming over for the Latin Christmas, which was the next day." I went down to the Mandelbaum Gate and saw the pilgrims come through and the like. I got an awfully nice reception at this consulate both in December and in June, so this is not the source of the trouble.

A few weeks later I went to Tel Aviv and, since we were going to be six months in the country, I had to register the family and take care of everything. I did this with the consulate. I was introduced to the consul; I think his name was Dorfmann [Stephen Palmer Dorsey? Ibid., 1954)], and he was extremely warm to us. But he was late coming in so I asked to see the security officer, and I met three or four people in the economic division of the consulate. I was very much at home there. I can't remember the name of the security officer of the consulate, but his initials were W.W.W. [William Woodard Walker, cons. Paris, Ibid., 1954, p. 512]. [He was real friendly and I teased him about McCarthy. [Roy] Cohn (1927-1986) and [G. David] Schine had been through this place a month or so earlier, and the hatred the Israelis had for these Jews was simply awful. We talked about McCarthyism a good deal.

Some weeks later I came back because I was giving a major speech, the first one to be given in Tel Aviv for the Israeli Association for the Advancement of Science. I was to be introduced by Ernest [David] Bergmann (1903-1975), who was the chief science advisor to [Premier David] Ben Gurion (1886-1973). Up to recently he'd been head of the Weizmann Institute of Science. I had invited the man in the information service and several people I thought were interested, to come hear my talk. It was the peaceful uses of atomic energy, again based on Eisenhower's speech, and they had given me copies of the speech in English and Hebrew. They'd also given me some wonderful posters they had in Hebrew about Atoms for Peace, which were appearing in barbershops and beauty parlors all over Israel. I sent some home as souvenirs and I distributed a couple more to the grammar school in Rehovot. [Julie attended fifth grade for part of the stay.]

The information officer here was a very decent person and he told me a lot more about the information service. I think this is a fine part of the United States government. I've also seen in Denmark what it means to the Danish people to have an American library. This information officer in Jerusalem was killed in an airplane accident about two years later. He had on his desk a major book on Syria by a Princeton historian Sitti (*sic*) [Philip Khuri Hitti (1896-1978], who is very anti-Israeli, and we talked. I'm very interested in probing out the Israeli-Arab differences, and the psychoses on both sides. There are darned few in Israel, but there are an awful lot of them on the Arab side. I felt a lot of rapport with the people in the [consulate]. When I went down to see them about my talk—anytime I give a prominent talk overseas, in this atmosphere, I want a member of the United States government as my personal guest. If you came and went with the Russian consuls, any arbitrary American later could say what you did. Oppenheimer went to dinner with a dangerous man, Haakon [Maurice] Chevalier (1901-1985), in Paris, with only his wife, who can't be an independent observer. If Oppenheimer had had any other arbitrary American present, he'd not have been in the trouble he was in. I sensed this in the Kamen case, and when I first met Kamen again in 1947 I had with me Richard Kenyon. I just felt that the first time I meet him I want to have somebody there who can testify to the spirit of the talk and the general details. I still will be this way. Here I have defenses, if you want to testify against me.

When I came back, I got a notice that this fellow who was the security officer wanted to see me. I went upstairs. He was away for 10 or 15 minutes and I sat around. He came in and said he was awfully sorry, but he had to give me a little lecture. He said, "I have in front of me a top secret telegram, and I can't tell you what it says but I can paraphrase it. You are being severely censured for the way you're running around Europe flashing AEC cards. It's a serious threat to security and I'm going to have to tell you to stop it." I said, "Well, what will I do? I'm asked to go to all these places and nobody knows what I should do, and it seems to me I have to identify myself." I said, "I've only shown this thing to American citizens who are officials. I haven't shown it to the Spanish and Egyptian and Israeli secretaries in the office." "Well," he said, "never mind, don't worry about it. I had to tell you this."

I found out in 1957 from my good friend, Ed Brady, who is now at the State Department, and who lived next door to one of the security men who had to worry about me, that the guys in Washington decided I was going to make fools of them by following the letter of the law and they thought I was trying to make a security issue by making all the trouble I could to conform to the obvious regulations.

I relaxed after this. I went back to Jordan and of course, I passed through the consulate. This is something you have to do when you pass this war line. I passed through the

American consulate in Jerusalem, where I was awfully nicely treated.

On the other hand, the consulate people were all against my getting rights for the PX [Post Exchange]. I thought that after all, if I have to report to the government, I ought to get some sort of privilege out of this thing. I'm in a sense an ambassador for the United States and I was working hard for American-Israeli science relations, and Egyptian-American when I had lectured in Cairo. I had also spent about $50 for telegrams and I billed the AEC for this after I returned, and didn't get it.

I also made some handwritten letters to Rolander, one of which I transmitted through Wilson. I made a 10- or 12-page analysis on the SS *Independence*, before I got to Gibraltar, of the security situation, because I had then decided on seeing Madame Joliot. Madame Joliot had just a few months earlier been denied admission to the American Chemical Society because it was claimed that she's a Communist. I'm sure she's not a member of the Communist Party. I wrote all this stuff to Rolander and I kept a carbon copy for myself. This wasn't military secrets, it was politics. I sent another handwritten letter through this security officer in Tel Aviv, and I guess I never sent any more.

In Cairo I called on the consulate. In Rome I just signed the guest book. I went to the consulate and found a big guest book, and I just signed my name, because I had relaxed. But when I got to Paris I felt quite differently. Paris was where Oppenheimer got into trouble with Chevalier, and all this stuff was now out. I decided not to see Haïssinsky, much as I love him, until I've seen someone at the Embassy, and whom

I'm going to see is obviously my dear friend, Jeffries Wyman. I went to the Embassy and found out he was away for the weekend, in Rome, and I thought I was told there was an AEC man in Paris. His name is [Howard A.] Robinson [Ibid., 1954, p. 422]. He's a physicist, but I believe that the story really is that he was a State Department specialist in atomic energy, in their office. There is now an AEC office in Paris, and I think George [F.] Quinn is the man who is presently there. I've seen him several times and have always been treated very cordially. Quinn doesn't act as if he thinks I am a total fool.

But Robinson was also away, so I had to wait a week. I arrived in Paris the 14th of July, Bastille Day, and the following Monday both men were due back. I figured if I didn't see Haïssinsky I couldn't see any of the French scientists, because they all know one another, and all of the radiochemists in Paris, irrespective of politics, are very warm to Haïssinsky. He's the dean of radiochemistry in Paris. He's a much more effective human being than Madame Joliot. She's too cold and distant, and Frédérick Joliot is really a physicist. So everybody loves Haïssinsky and sees him around, and has him to cocktail parties, and gets his advice.

I was a little bit angry. I wanted to have a little science contact, but I decided to talk to Wyman first. Wyman was glad to see me. He said, "Charles, you're an idiot. Don't worry about any of this stuff, of course you can see these people." Then he told me how angry he was at the State Department, how they treated Haïssinsky for the visa story. Haïssinsky was dragged before the United States Consul when he wanted a visa in 1949, and he was asked to specify how he voted in every French election since 1943. He was asked to detail his reactions to the Korean War. Now this makes me boil. The United States government can't ask us how we vote or what religion we belong to, and how can United States officials ask eminent foreigners who are being invited to the United States things they're denied the right to ask us?

Then he told me the story of the Haakon Chevalier dinner, and this I'll put in the record even though it's a secondhand story. I don't know where Jeffries Wyman is now. He left the State Department after his normal two years, was back at Harvard for a year, and then became UNESCO representative in Cairo for all the Middle East, including Israel. It turned out UNESCO cleverly dropped Israel out to save trouble with the Arabs, but he felt as I did that the thing [to do] was to foster scientific contacts between the Israelis and the Arabs. I was in fairly good correspondence with him until about a year and a half ago. But he just loves remote parts of the world—the water tribes of Burma. He went to Korea just before the Korean War broke out, but he couldn't get there because of the war and he was in Japan. He's sort of an amateur anthropologist. His second [third] wife is a Russian girl, with anthropology, cultural history leanings. I met her in the United States about 1958.

Wyman told me that Oppenheimer had done a very great service to the United States in talking with Chevalier. Chevalier had had his passport removed. He was an UNESCO employee. He'd been a former member of the English department at Berkeley, and I think when Oppenheimer married his wife, Kitty, she was very unpopular at Berkeley and the Chevaliers were two of the closest friends they had. Wyman said little about Chevalier's approaching Oppenheimer in the kitchen at Berkeley, but I suspect this wasn't a professional spy approach. It was just why the hell can't we do something about letting the Russians know, sort of the feeling Kamen had—let's swap information for information.

Wyman's job was to find out what was wrong with Chevalier, to protect the State Department. Nobody knew. The removal of the passport was a Department of Justice affair. The State Department didn't know what was going on, and they were sort of unhappy about it. He wanted to send information as to what in the hell Chevalier might do next, to save damaging United States-UNESCO relations, and the Oppenheimer-Chevalier lunch had served a very valuable purpose, personally, to Wyman. Wyman had been a graduate student at Cambridge with Oppenheimer, and knew him very well and thought very highly of him.

Wyman says that when the Oppenheimer case broke he was in Rome on State Department business and a telegram came from Washington that a special emissary from the United States was coming to interview him privately and was due on a 10 p.m. plane. Would he meet him and save the evening for him?

The plane didn't come in but Wyman stayed up till about 2 o'clock and met him the next day. Wyman and this State Department messenger spent about six hours together, and this man just combed every single contact and every single thought Wyman had about Oppenheimer. The hearings were going on and the State Department wanted to know really what happened. I take it he was a State Department emissary, and I doubt this thing ever got to the Gray Committee. But the State Department had to protect itself against Strauss and McCarthy.

Wyman was a bitter anti-McCarthy man, and he laughed about the Cohn and Schine story and told me some more. I was about an hour with him. I told him the reason I hadn't seen Haïssinsky was because I didn't want to be caught in an Oppenheimer sort of affair. We talked about this Chevalier dinner very, very much because it had worried me a good deal. I had guessed some of these things. I never met Kitty Oppenheimer and I don't want to be in any way bearer of derogatory information, but I know there were some emotional tensions in the Caltech and Berkeley community about friends of her second husband, and just other relations involving the two Oppenheimers.

Then I went to see Robinson whom I'd never met before, and I told him I was amazed and pleased to find *Physical Review* lining the wall of his office at the embassy. This was a fine thing for Paris. I told him I had been instructed to report to some United States officer, and while I had just talked to Wyman, and since I had atomic energy relations and he did, too, I thought I ought to report to him. But no officer I reported to had the slightest idea what to do with me. He said, "Don't worry. This is a piece of utter nonsense. The man waiting to see me after you is the president of a small steel concern who has been told the same thing, and he's shuddering for fright that he might be beat over the head by a Communist as soon as he leaves this place. People are scared to death of Communism in Paris." I told him a little— we talked a little about the Chevalier case, and a good deal more about what I thought of Israel and I told him what I'd done in Israel, and I left. I don't think I've seen Robinson since.

The rest of the tour, when I went to Belgium and the Netherlands, I don't think I did a damn thing about the United States. I was sick and tired and I figured that my talk with Wyman had settled everything. Wyman felt that, in spite of the Chevalier case, if they were out to get Oppenheimer it didn't matter what Oppenheimer had done in Paris. But it is true that the Chevalier dinner in Paris is the reason Eugene Zuckert gave for voting with Strauss rather than with Smyth.

I know Zuckert awfully well, though I have never been alone with him to talk about this. He's chairman of the board of directors of Nuclear Science and Engineering Corporation of Pittsburgh, and I'm the most outspoken member of the board. We're almost always in accord, but we're busy on company matters the two or three hours we see each other almost every month. I'm supposed to see him September 12th for a board meeting, and he wants to talk to me ahead of time about some company problems. I'll most certainly quiz him on this particular point.

Zuckert was also a very close friend of Gordon Dean's, and was promoting him for President of the United States as a dark-horse candidate for the Democratic Party in 1960, though Dean was killed in an airplane accident August 15th [1958] in Nantucket.

Well, this was my view of the Oppenheimer situation as seen from Europe, and with the certainty I was involved in these hearings. Let's finish the Oppenheimer side and then I'll tell about these hearings. I think the hearings are pertinent also, more than the Pauling case, which is public knowledge to a high degree. It isn't finished, either, and won't be for a long time.

On coming home, I read everything I could lay hands on. I got home September 20th. Late September was probably the [Senator Ralph Edward] Flanders (1880-1970) speech in the Senate attacking McCarthy and calling for the censure. I'm very glad I was out of the United States when all this TV program was going on about the McCarthy hearings because of the waste of time involved. The country was fascinated. Anywhere I went and people found I was an atomic scientist, without exception—trolley cars or anywhere—people would say, "What do you think of Oppenheimer?" They were amazed when I said, "I know Oppenheimer fairly well and I love him and I think he's a great man and I think this is criminal what happened." "Well, what about Teller?" "I know Teller lots better, and I think he is one of the finest scientists I've ever known, but I think he's an idiot in politics."

The American public was aroused by the Oppenheimer case all through society. I was glad to talk about it when I realized, "Here's a man from Mars, seeing the case from the outside, but I know personally two of the protagonists." I also met Strauss very briefly in October 1954. He was president of the Tracerlab conference at the Statler Hotel in Boston. He gave a good talk, as a matter of fact.

Lewis Strauss is a person of very high ability, but he's crooked as hell. Honest-to-god crooked, in some things. And he has this awful emotional distortion that his way is right come hell or high water, and he'll break heads, hearts, jobs, or reputations unless you see things his way. I think it's a very good thing he was denied the right to be Secretary of Commerce and I'm proud of the fact it was Dave English, a very modest, conservative physicist from the Argonne National Laboratory, then chairman of the American Federation of Scientists, who had the courage to oppose Strauss for Secretary of Commerce in November 1958. English was a little inept in how he did it, because it was David attacking Goliath. A lot of people were unhappy, but the council of the Federation had voted, 24 to 0, with one abstention, for the first time in history, to enter the lists on a personality case against a man. I think the FAS did the right thing. I think the FAS lost a little face, but about a month later the thing the FAS was fighting in Strauss caught up with him. The *Washington Post* turned on Strauss when it was found out he was doing in the Commerce Department the same sort of dirty thing, giving half-truths and avoiding giving necessary information to people, justifying his decisions, which he'd always done in the AEC and which make him emotionally and constitutionally unsuitable to be a chairman.

I must say on behalf of Strauss that very many very good things were done in atomic energy involving business developments of atomic energy under special pressure from him and, to the best of my knowledge, no one else in the Commission. I also know Willard F. Libby rather well, and saw him several times in the fall of 1954, and feel he was a tool of Strauss by the accident of having no more political insight than Eisenhower had when he came into office, and he felt that as a Republican he had a sworn public position and the Atomic Energy Commission, for the first time, was partisan. Strauss felt that party line counted and Dixon-Yates [TVA power plant controversy] was coming up. I told Libby when he asked me about Dixon-Yates, "All I know is what I read in the *New Republic,* and I think it stinks. The safest thing is to stay out of it. It's a terribly complicated power struggle." Libby was very frightened in November 1954. Senator Kefauver just was massacring Libby at the hearings for Libby's appointment to the AEC because he was striking at Strauss through Libby. Libby passed, as he obviously should. He's a very able scientist and I think he may be the one who shares the Nobel Prize with Kamen for carbon-14— Kamen for discovering it and Libby for getting it in nature. But I think Libby is an idiot in Washington; he's a wonderful man at a university. I'm glad he's at UCLA, my old home base. And I wouldn't mind if Libby heard these statements about him.

I phoned Libby on Thanksgiving Day, 1955, to get him to participate in a scheme for the United States to give two reactors to Palestine, power reactors to 5- or 10,000 kilowatts to pump water to the city of Jerusalem. It would have been a wonderful Christmas gift from the United States government, or atoms for peace program, to a power plant on the border between Israel and Jordan, a very troubled border, one on the Israeli side to pump water from sea level at Tel Aviv to Israeli Jerusalem, and one on the Jordanian side—an identical one—to pump water from the Jordan, 1,300 feet below sea level, to the Arab side. Both sides of the city suffered from lack of water and power. One-third of the power consumption in Israel is water-pumping and a good fraction of this is pumping water for Jerusalem because they've got to keep the city going as an emotional symbol, and the Arabs feel the same way. In Arab Jerusalem you have water every other day in most parts of the city. Hotels have tanks on the roof to give people, but they ask you to conserve water and they're very careful the plumbing doesn't leak because they can't afford the waste.

Well, this was the campaign I had in the fall of 1954, and Gordon Dean was rather prominently involved. I phoned Libby to tell him I was leaving for Chicago, and going to see James Franck, Sam Allison, and Harold Urey. Urey's one of my gods if he wasn't mentioned before. That means I was promoting him to a god and Dean joined my circle of gods at this time, because he was superb. Libby was interested and turned out to be very helpful in this thing. He, too, felt that the present need is some quick symbol. The IAEA [International Atomic Energy Agency] was too slow coming, and the power programs for the other nations—the other nations were too slow accepting them. As a matter of fact, I think only four nations have reactor money from the United States in their hand. Denmark got it the other day. Twenty nations have signed contracts but they just haven't the skill to use it yet. Portugal's halfway done, seven years later. But something [done] quickly and constructive, like something to lower the tensions in Palestine, would have been a good thing.

Libby said he couldn't come to the American Physical Society, which was my technical reason for going to Chicago, because he was staying home studying power engineering. He'd been asked by Senator Kefauver how could he be on the Atomic Energy Commission when he didn't know anything about power, whereas Libby is, without question, even better than Seaborg as a radiochemist. He's without question the best radiochemist in the country in its history, in my opinion, though Fermi felt that Seaborg was the boy. Libby was scared to death that when the Johnson-Neumann [bill] came up for hearing Monday, he'd be massacred, too. I said, "Bill Libby, don't be an idiot. You're attacked because you're an American, and it's easy for Congress to get headlines out of you. Kefauver wants some headlines. He's running for President and he has to make the papers. But they're going to be very decent to a distinguished foreigner who's become an American citizen and who has an accent. You're the symbol of politics, not he." And it turned out to be that way.

I talked to Gordon Dean by telephone about this time, and I guess I saw him. Dean says, "The answer is simple. Just quietly write to Kefauver." Gordon Dean telephoned Kefauver. He said, "For God's sake, you've had your fireworks on Libby. Lay off and let the guy in. You've got to go ahead with Atomic Energy Commission work." I wrote a friendly letter to Kefauver and got a nice answer.

I was talking about the atmosphere in the United States in the fall of 1954. The story that friends told me was that Strauss knew that McCarthy was going to attack Oppenheimer, and Strauss refused to be in a position of defending Oppenheimer. He hated him. I'm not surprised. I'm sure that Oppenheimer was as arrogant to Strauss as Strauss was to Oppenheimer in the isotopes for overseas. I had more trouble understanding the role Teller played, but the newspapers said that he was very jealous that he wasn't the Father of the Atom Bomb and this was because Army security kept him out at the critical time. He therefore would try to be the Father of the Hydrogen Bomb.

He had, of course, opposed Oppenheimer back in 1951, and in a sense Teller was more right than Oppenheimer. Then, there was also this little mess that Oppenheimer had misquoted what Seaborg had said. This may be simple forgetting. Seaborg had said he couldn't vote with the majority, and Oppenheimer didn't tell this to the others, so that the members then present were uniformly against it. I also found out that Oppenheimer had planned a meeting in February, 1951, about the hydrogen bombs, where they were going to get about 100 physicists at Princeton, because Oppenheimer felt that you get the answers to a problem not by a small committee with its biases and its leaders, but you have to have a sort of a democratic fight. You'd have Teller, and Oppenheimer would try to stay out of it as arbiter, and you'd have guys like Sam Allison—I don't know whether your father may have been involved in this thing. But if you have about 100 people staying about a week, with total access to all information—all these people were totally cleared and the like—out of this would come the best thing that scientists could see for the United States in 1951. I expect at this meeting there should have been the military men and all the others. There certainly would be the Atomic Energy Commissioners.

When President Truman made his unilateral decision on January 31, 1951, this meeting was without purpose, and it was cancelled as it should have been. Oppenheimer, as soon as the thing was made official by the President, went along with the program. He did what he could. I knew these things were involved, although I hadn't known about this meeting till I came back in the fall of 1954.

I also read most of the Oppenheimer hearings as fast as I could, though I never had time to come back and read the rest. I think that the Oppenheimer case ranks in the world about the same level as the Dreyfus case. The Dreyfus case was Semitism versus anti-Semitism, this is Jew against Jew against Jew. It is a Jewish fight which has the same nastiness as the fights in Israel have, unforgiving forever. The fights between Jews often extend between Jewish students, whereas Christian fighters like Seaborg and Coryell or Fieser and Woodward at Harvard are real big, noisy fights but they don't go down to the next generation. Seaborg's been wonderful to my young men. He has two of them now and his most recent PhD is instructor at MIT. The fight's over, as far as impeding science, but it was as dirty as you can get when it was in the middle.

It also seemed that a good deal of the driving force for the Oppenheimer fight was fear of the massive retaliation, preventive war gang in the Air Force, of scientists messing around in Defense and spending fifty billion dollars on the air defense of the United States. This was the estimated cost of the alliance we now have, which probably cost about eight billion dollars, because radar got better. The fear that university scientists are softheaded and some of them are disloyal. The thing to do is attack the universities, but they are awfully hard to attack. Universities have a great position and love with the American people. But individual professors could be made idiots, and it was decided Oppenheimer was a weak link, partly through his political promiscuity in the late thirties, which I think is due to the fact that his understanding of what was going on in Germany was an anti-Semitic thing. He didn't assume the responsibility of the Jew to worry about this, he never worried about Jewishness before that. Secondly, because his wife was politically vulnerable and also probably had enemies who had helped pass bad stories about her without the corresponding good ones.

I also felt that the high tide of the anti-intellectualism of McCarthy came into its peak in the Oppenheimer hearings. The Gray Committee, it seems to me, violated every principle of the Atomic Energy Commission with this son of a bitch [Roger] Robb, who was a prosecuting attorney, and the bill of particulars of another cold-blooded bastard, Nichols. [McMillan, 2005.]

Strauss has a reasonably good position in my mind with relation to Oppenheimer [blank] something of the world reaction to the Oppenheimer [blank]. I think he played a dirty, crooked trick, publishing that without any editing except his own. It has security errors in it. I'd love to have him nailed on these just for poetic justice, because he used security against people in political affairs often. The silliest thing is making a wall of secrecy around Oppenheimer. You cut off a productive source of ideas for the benefit of your enemies, not the United States. This is fantastic. It still stands, you know. Eisenhower doesn't have the intelligence or the courage to reverse it.

Anyway, Strauss brought this thing before the Atomic Energy Commission and reversed the Gray Committee position that Oppenheimer was disloyal. Strauss engineered and wrote the majority position that Oppenheimer cannot have his security [clearance]. He had to protect himself administratively by saying that Oppenheimer hadn't the proper character. These may well be, in Strauss' eyes and a lot of other people's eyes.

But then Strauss went to a meeting of the Board of Directors of the Institute for Advanced Studies at Princeton (this is now October, 1955), and had the courage to vote for Oppenheimer when he could have abstained or been absent, and had the courage to tell the press as he left the room. He left the conference early. This I consider is as much apology as the world will ever see from Lewis Strauss, but I still think it's a substantial apology—

JBS: You mean voting for Oppenheimer as head of the Institute?

Coryell: He voted to renew the term. But I think that the fact that Strauss voted with Oppenheimer and told the press voluntarily before the meeting that he had, is as close to an apology as we'll get. I think somebody smart could have reversed this clearance problem, but I think it is also an academic problem now.

I felt already with the attack of Flanders on McCarthy, the fact that McCarthy was kept out of the Oppenheimer case— probably by clever manipulation of Strauss. It could have been much messier, because I think Strauss has sense and I don't think McCarthy did—the tide of anti-intellectualism reached its peak in the Gray Committee and Strauss already helped reverse it by mitigating this. I'm just sorry that Zuckert misinterpreted the Chevalier dinner, which he said was the only reason he voted against Oppenheimer—this was dangerous and a sign of disloyalty to curry favor with a man who's a pariah of the United States. Perhaps what I told you from Wyman is valuable new information on this thing, but you'd better get it from Wyman himself.

The Flanders attack was wonderful. Then the vote on the censure of McCarthy. Two of the three censure movements passed, you know, though didn't he get named something besides censure? McCarthyism rapidly fell back. I was very glad when McCarthy died, but I was glad he was defeated before he died. He's a hero now for about 25% of the American people. I'm sorry that [John Fitzgerald] Kennedy (1917-1963) never came out publicly against McCarthy, but I think I understand—

JBS: In 1956, he did.

Coryell: No, the point is—

JBS: You mean, at the time.

Coryell: Yes. This is what Eleanor Roosevelt can't hold against him. I think that Kennedy was a really sick person in Florida, and I'm sure that people around Kennedy in those days were mostly Irishmen, who unconsciously feel that McCarthy is one of their kind and that Communism is god-awful. No matter if he has bad manners; he helps. They therefore figured that good politics is not to speak your mind. I wrote Senator Saltonstall. I don't remember writing Kennedy, but I must have. Saltonstall wrote back, just before the censure, "The Senate is a jury to decide the McCarthy case. I cannot commit myself ahead of time." There was said to be an understanding in Massachusetts that neither side use McCarthy. The Irishman could use McCarthy, but the Democrats and Republicans weren't sure what effect it would have on politics and I think both sides felt that it was dangerous. I think the decision not to use McCarthy in the state is somewhat to the credit of the candidate of each party.

I wrote my feelings about this thing in some detail to Edward Teller, for whom I have lots of affection still, though I haven't seen him to speak to since 1951 in Los Alamos, when he told me there was a crisis on and in the case of a national crisis would I come to Los Alamos and work for him, and I said yes. I wrote a long letter to Teller and told him I wasn't going to show this letter to anybody. I gave my then-analysis of the Oppenheimer case, which is roughly what I say now. This was an attack on the university system, and all, as the Blair and Shepley book gives ample evidence. This is wonderful proof of a good deal of my insight, but it's a question of seeing facts the same way and interpreting them opposite. I was begging some sort of a public apology, or some sort of public recognition that Oppenheimer had done some good and this wasn't so bad from Teller and Strauss. I said that I hoped he would send a copy of this letter to Strauss.

I got a nice letter from Teller about ten days later. This was about mid-October, 1955. I think it was before the McCarthy censure. The letter from Teller I'll find some day, my own letter I think I destroyed. Teller wrote back and thanked me very much for my warm letter. He said, "I think you're substantially right. I do, however, feel that McCarthy did a lot of good in spite of the evil he did." I was also told about this time that E. O. Lawrence was very strongly pro-McCarthy, and so was [Luis Alvarez]. And Berkeley was the only university that Oppenheimer didn't win a moral victory in. In all the government labs involving atomic energy, I think Oppenheimer had a clear moral victory.

I think if you can stand it I'll tell you the story of the Goldring-Rosenbluth hearings.

I'll say the last few things I want to say about Strauss. I mentioned I met him briefly in October 1954, in Boston. I hoped to develop the contact I'd had in 1948 in correspondence with him and have a chance to talk privately with him and see if Teller had ever mentioned the matter to him and talk over my point of view. My point of view, in 1954, was, as it is now, to make peace on the Oppenheimer case by some reasonably decent apology to Oppenheimer of the—I think that Strauss has done enough of it. It's the government, it's Eisenhower, or Nixon's turn—and go on with constructive things. It was a great butch, and there's no kidding anybody who has any sensitivity about science when they read this thing.

I thought I could outsmart Strauss and capture his interest again in an unusual way. I brought my only copy in Hebrew of President Eisenhower's speech on atoms for peace. Strauss is said to be, according to *Life*, a conservative Jew, a devout man who reads the Scripture in the synagogue every Friday night. I thought he would be delighted. Most Jews love Israel and love the Hebrew language. This thing is a backwards thing like an Israeli thing, with a lovely picture of Eisenhower and not a word of English in the whole thing, and it gives atoms for peace. *Shalom* is the Hebrew word for peace, and *atoms* is the Hebrew word for atoms, and if Strauss is any good as a Jew and reads the Torah, he can read Hebrew. As a matter of fact, anybody that reads Yiddish can read Hebrew. [With command of the alphabets.

Before going to Israel CDC averred that it took him 70 hours to learn 70 words of Hebrew, the lowest intellectual yield of his life.]

I handed this to him as he was leaving after his speech. I said, "I'd like to have a chance to talk with you. I'm Charles Coryell; I wrote you in 1948." He said, "Oh, I'm glad to meet you, Dr. Coryell. I have to catch a plane in about ten minutes." I said, "Here's a little souvenir." He said, "What's this?" I said, "This is President Eisenhower's atoms for peace speech translated into Hebrew. I picked it up in Tel Aviv. I thought you'd like it for a souvenir." He said, "I can't read this. Tell me what it says." I said, "Atoms for peace." I gave him the Hebrew words and the English. This is just a plain, crooked deal on him to say that he's a Jew or emphasizing it at the right time. He never did respond. I'm not sure he remembers my name. I did give it. But he may also have felt I was involved in the Oppenheimer case and he'd better stay clear of all emotional scientists because there's nothing but trouble in them for him.

Sometime in 1956 or 1957, Grace Mary and I were guests of the governor of Maine. Before this we were guests of Jim Reed. We met at a cocktail party Sumner [T.] Pike (1891-1976), whom I'd known for some time. I brought Grace Mary to this cocktail party in Maine, having to do with Maine atomic energy developments. Pike said, "Oh, I remember you, Charlie Coryell. You're one of the guys that's mad at Strauss." [Few called Charles Charlie.] And he turned to a whole bunch of people, and he said, "That son of a bitch Strauss is one of the worst crooks I've ever seen. I've seen that man take the tape recording of a man's testimony and edit it a year later to reverse the meaning, and play it back to the man, saying, 'See, this is what you said'."

Epilogue

The narrative trails off. Joan Safford explained that she and Charles discussed one last date, before she and Pancho, Franklin Safford, had to return to New York by September 7, 1960. The next day, Monday, she bought chicken at a Cambridge grocery store with the result that she and Pancho both developed food poisoning. With the help of family to clean the home they housesat, they made their deadline, but Joan has never bought chicken on a Monday since. That summer she interviewed first her father, Kenneth T. Bainbridge, then Norman Ramsey, and third, Charles and typed the first two transcripts and first three sessions with Charles herself on her portable Olivetti Lettera 22. She was paid $15 per contact hour. That fall she left Columbia oral research to work on the Kennedy presidential campaign.

The narrative essentially ends with the Oppenheimer case of 1954. William Walters has described that in 1956, Charles was first to interpret the peaks in Harold Urey's solar-system abundance data as arising from *two* processes occurring at closed neutron shells during the nucleosynthesis of elements in stars. One was later named the s-process for "slow" that goes on in red giant stars continuously. The other was named the r-process for "rapid" that is thought to take place in a few seconds during a Type II supernova. (Walters, ORNL, May 2007). David Freeman, then Charles' graduate student remembers the seminar when Charles saw evidence in Urey's discussion and interrupted him. "Urey was vigorously less certain. Seminars at MIT rarely involved a passive audience, but this scene was the opposite." (Email, December 27, 2011).

Charles experienced a boom then. He wrote to daughter Patty how he longed to generate ideas without the burden of footnoting and proofreading and did not publish, but noted his work in his May 1956 Annual Report (MIT-LNS, AECU-3379). A reviewer, probably Harold Urey, added Charles' name as the 13th footnote to the germinal paper by F. Hoyle, William A. Fowler, G. R. Burbidge and E. M. Burbidge, "Origin of the Elements in Stars" (*Science*, 124:3223, 5 October 1956). They enlarged the paper, known for the initials of the authors, B²FH, "Synthesis of Elements in Stars" (*Reviews of Modern Physics*, 29, 47, 1957), in Fowler's words, "in which we showed that all of the elements from carbon to uranium could be produced by nuclear processes in stars starting with the hydrogen and helium produced in the big bang" (W. A. Fowler, Nobel lecture, December 8, 1983, www.nobelprize.org).

Charles spent the first half of 1963 on sabbatical with then Director Moïse Haïssinsky, at the Institut du Radium, Sorbonne, Paris, gamely challenging himself in French. In the spring his brother, William Harlan Coryell, Jr., one of four crewmembers of a Flying Tigers cargo plane, died in a fire at San Francisco Airport. His psychological cycle deepened, culminating in his first hospitalization the weekend of the Kennedy assassination. During the following week Professor Glen Gordon on Charles' behalf hosted William Walters as a candidate for a faculty position in nuclear chemistry. Although hired, not long afterwards Professor F. Albert Cotton catalysed a shift in the Department away from radiochemistry to molecular biochemistry. In the fourth year of his battle with cancer, by refusing tenure to Professor Gordon, MIT moved to close Charles' Laboratory for Nuclear Science, which he affectionately named Arthur A. Noyes Nuclear Chemistry Center $A^2N^2C^2$. Professors Gordon and Walters, with graduate students, post-docs, and much equipment moved to the University of Maryland.

Deserving of understanding is the issue of treatment of creative academicians who experience periods of mental illness. The turmoil affects self, family, and colleagues. The works of Kay Redfield Jamison on manic depression are illuminating and compassionate (1996, 2000, 2005, 2007). The persistence of manic depression in human genetics may lend a protection against failure, an ability to recover and initiate new ventures.

In his memoir, *Adventures in the Atomic Age,* (G. T. Seaborg and E. Seaborg, 2001), Glenn Seaborg describes his ailing from the workload in 1942 at the Metallurgical Lab at Chicago. To break his lingering flu symptoms, he took up golf, and later hiking. He lived long and productively. Charles embraced first a central love of science and high standard of personal integrity enhanced by the examples of Noyes, Pauling, Franck, and "his circle of gods." Legatees might consider the benefits of healthy habits of self-care, balance, even in the face of worst fears. Charles recorded in Session Five, "I knew that the timing of the job and the importance of the Project were immense, and I literally feel I was within a short time of being a bottleneck of the atomic bomb." The work schooled him to an unusually intense work ethic.

As for his theme of political orientation for scientists, Charles continued to work for public understanding of atomic energy, peaceful uses, and an end to war. In his first year at MIT he gave 65 public lectures. He participated in at least two Pugwash Conferences, the Committee for a Sane Nuclear Policy, Federation of American Scientists. Subscriptions to *The New Yorker, Bulletin of Atomic Scientists, I. F. Stone's Weekly, The New Republic, Newsweek, Life,* piled up in our home. He carried on extensive correspondence with carbon copies, and, to Grace Mary's chagrin, turned his suit pockets into portable filing boxes of telegrams, letters, and index cards.

After Seelye Martin and I married in the Unitarian-Universalist Follen Community Church, Lexington, June 24, 1965, and after his right leg mid-thigh amputation for giant-cell bone sarcoma in April 1966, Charles joined that church. He often stood one-legged on crutches in vigils against the Vietnam War. Toward his cancer, which shifted between hard "golf-ball-sized bone tumors and aspiratable synovial fluid sacs that filtered into his lungs, he declared, "a rare cancer has no statistical basis, so I will live until I die." He took a scientific approach and particularly welcomed the partnership in treatments offered at the Pasadena Tumor Institute in California. He noted with curiosity that his tumors accumulated iron and pondered returning to biochemistry after his post-doctoral research with Linus Pauling on the magnetic properties of the heme of hemoglobin. He traveled solo to Israel and to Turkey. He married Barbara Ogilvie Buchman in 1969. On their honeymoon trip to the Soviet Union and her childhood home of Irkutsk, where her father's eye clinic was still operating, his address book was stolen. He recreated it by hand, but also kept a record of his temperature on the backs of the entries. Mornings he felt better and worked. Afternoons when the fever rose, he rested on the sofa, read, and talked on the phone.

He delighted in the Atomic Energy Commission Award bestowed by Glenn T. Seaborg on March 4, 1970. In the fall when the cancer cells invaded his last two remaining lung lobes he helped me face the truth. I had not been able to admit my fear of losing him. Gently he said, "You have reason to feel angry with me for imposing on you." And, "I do not fear my own dying, I feel I have lived three lives in one." He died in Deaconess Hospital, Boston, the evening of January 7, 1971.

The last words of Charles' *Reminiscences*: "See, this is what you said" apply for Charles, too. Chemists will know better than daughters how his creativity flourished, inspired, propagated. Still, his words, his honesty, his prodigious memory, his teaching, his love, his ebullience animate our memories, collegiality, and friendships.

Lexington, Massachusetts, May 21, 2012

Illustrations

Coryell Fission Products Group,
Oak Ridge, October 1945

Pictured clockwise from left: (standing) Warren Burgus, Lionel Goldring, Charles Stanley, Jack Siegel, Russ Williams, Ed Shapiro, Howard Gest, Ray Edwards, Don Schover, Dick Money, Richard K. Bersohn, Jack Marinsky, Lawrence Glendenin, June Babbitt, Ed Brady (center front), Gerald Strickland, Henri Levy.

Others in Coryell-Sugarman Groups recalled by CDC in 1962: B. M. Abraham, D. S. Ballantine, G. W. Campbell, C. R. Dillard, N. Elliott, H. B. Evans, B. J. Finkle, M. S. Freedman, A. Goldstein, E. J. Hoagland, S. Katcoff, G. R. Leader,D. C. Lincoln, R. P. Metcalf, T. B. Novey, E. R. Purchase, W. Rubinson, C. Schwob, B. Selikson, L. G. Stang, L. Winsberg, R. M. Adams, N. E. Ballou, T. H. Davies, D. W. Engelkemeir, M. H. Feldman, H. Finston, M. B. Hawkins, D. N. Hume, J. D. Knight, C.M. Nelson, G. W. Parker, S. Raynor, B. Schloss, R. P. Schuman, J. A. Seiler, E. P. Steinberg, A. Turkevich, E. J. Young.

ODE TO A LOST NICKEL

A nickel is a piece of dough
Whose value can't be questioned
Since Uncle Sam defines it so
You gotta take it, even though
You know darn well, as alloys go,
Intrinsically it's less than.

Now this same nickel can become
A symbol strong and mighty
When harnessed to a proper bet
Involving jobs that must be met
By Coryell and his whole set
Of chemists, bright but flighty.

We lost the bet - we're *glad* we lost
And so, we judge, is Oppie
Who, wrapped in darkness and in doubt
Now knowing what 'twas all about
But sure that it would not get out
Had almost blown his toppie.

So sing the praises of the coin
And sing also of chemists.
Who break the rules and spoil the view
And yet when pressed with something new
They make a bet and then come through
With much more than the limits.

> R. L. Doan
> September 18, 1944
> Barium day, Oak Ridge

Street View

Back Yard

Living Room.

The Coryell Cemesto "B" House at 137 Outer Drive,
Oak Ridge, Tennessee

A PICTURE

On my mantel is a picture of the atomic bomb
Taken in the first second after the first explosion.
I don't quite know where to put it, though it's been there
Nearly a year.
It attracts the attention
Of everyone who comes into my house.

Perhaps it is just as well that I can't quite find a place for it.
Everyone reads the personal note on the back of the picture.
Once the comment was made about its
                              similarity to a Japanese print.

What should I do if an English friend
Or a Russian, Chinese, French or Spanish friend
Came into my house?

That can't happen just now because I still live
In a restricted area.
But suppose they did come
Should I hide it?

I can't take down one picture and replace it
With my photograph of the bomb.
I can't throw it out
Because it has a personal note on the back of it
To me.

It is on the mantel like a cablegram which has just arrived.
Nearly a year is a long time for it to stay unanswered.

This is a new thing
Which came into my house and my thoughts
Just after everything else was in place.

The coming of the atomic bomb to the world
Has shattered all thoughts and ideas of the past
As dramatically as it crystallized the sand of New Mexico
Into a film of pale green glass.

And I don't know where to put it.
Neither do the Army and the Navy, or the U.N. or the Russians.

The rocket bomb does not move slowly like thoughts in the past
It takes into consideration and into calculation
The curvature of the earth and the winds of the stratosphere.

My thinking must be geared to rocket speed
And to the size of the earth.
The rocket and the atomic bomb can come
                              from any land overseas
To destroy the mantel where I have my picture.

You have your own fireplace and a mantel.
And, you, too, have a picture.
If you reflect, you will find
It bears a note
Which is very personal.

                    Grace Mary Seeley Coryell

Grace Mary Seeley Coryell,
Lexington, Massachusetts, 1947.

With the signed inscription, "Best regards to
Charles and Grace Mary, Nathan Sugarman and
Anthony Turkevich," and typed description, "This
photograph was taken by Jack Aeby at the test site
of the 'Atomic' bomb. Alamagordo, New Mexico,
July 16, 1945," the gift inspired her poem.

Julie E. Coryell with Promethium, 1947

Charles D. Coryell at M.I.T. with three-volume
Plutonium Project Report (foreground) ~1955

Charles D. Coryell thinking ~1951
Ben A. Shaver, MIT, photographer

My father, Charles D. Coryell, as I remember him - JEC
Ben A. Shaver, MIT, photographer

Charles D. Coryell with Pm co-discoverers:
Lawrence E. Glendenin and Jacob A. Marinsky

Barbara Buchman Coryell (third wife), Charles D. Coryell,
Patricia L. Huber (first daughter), Glenn T. Seaborg,
Atomic Energy Commission Award Ceremony, MIT, 1970

Selected Bibliography

by Julie E. Coryell

Bernstein, Jeremy, *Hans Bethe, Prophet of Energy*, New York: Basic Books, Inc., 1980.

_____, *Oppenheimer*, Chicago: Ivan R. Dee, 2004.

_____, *Plutonium: A History of the World's Most Dangerous Element*, Washington, D. C.: Joseph Henry Press, 2007.

Biographic Register of the Department of State, Washington, D. C.: U. S. Government Printing Office, 1953 and 1954.

Bird, Kai, *Crossing Mandelbaum Gate: Coming of Age Between the Arabs and the Israelis, 1956-1978*, New York: Scribner, 2010.

Christy, Robert F., Interview by Sara Lippincott, June 1994. http://oralhistories.library.caltech.edu/129/

*Compton, Arthur Holly, *Atomic Quest: A Personal Narrative*, New York: Oxford University Press, 1956.

Conant, Jennet, *109 East Palace: Robert Oppenheimer and the Secret City of Los Alamos*, New York: Simon & Schuster, 2005.

Cooking Behind the Fence: Recipes and Recollections from the Oak Ridge '43 Club, Oak Ridge, Tenn.: Oak Ridge Heritage and Preservation Association, 2003.

*Coryell, C. D. and N. Sugarman, editors, *Radiochemical Studies: The Fission Products*, New York: McGraw Hill, 1951.

*Fermi, Laura, *Atoms in the Family: My Life with Enrico Fermi*, Chicago: The University of Chicago Press, 1954.

Fermi, Rachel, and Esther Samra, *Photographs from the Secret World of the Manhattan Project*, New York: Harry N. Abrams, Inc., 1995.

Foerstel, Herbert N., *Secret Science: Federal Control of American Science and Technology*, Westport, Conn.: Praeger, 1993, 49-96.

Genung, R. K., R. L. Jolley, and J. E. Mrochek, *A Brief History of the Chemical Technology Division*, ORNL/M-2733, Oak Ridge, Tenn.: Oak Ridge National Laboratory, May 1993, 1.1 – 2.5.

**Gest, Howard, "The July 1945 Szilard Petition on the Atomic Bomb, Memoir by a Signer in Oak Ridge," 1995. http://sites.bio.indiana.edu/~gest/hgSzilard.pdf

Gibney, Frank, ed., and Beth Cary, tr., *Sensō: The Japanese Remember the Pacific War (Letters to the Editor of Asahi Shimbun)*, New York: An East Gate Book, 1995.

Gordin, Michael D., *Five Days in August: How World War II Became a Nuclear War*, Princeton: Princeton University Press, 2007.

Gordon, Glen E., "Obituary, Charles DuBois Coryell," *Journal of Inorganic Nuclear Chemistry*, Vol. 34, pp. 1-11, 1972. [Lists doctoral theses and papers.]

Selected Bibliography

Grinstein, Louise S., R. K. Rose, and M. H. Rafailovich, eds., *Women in Chemistry and Physics: A Bibliographic Sourcebook*, Westport, Conn.: Greenwood Press, 1993.

*Groves, Leslie R., *Now It Can Be Told: The Story of the Manhattan Project*, New York: Harper, 1962.

Hager, Thomas, *Force of Nature: The Life of Linus Pauling*, New York: Simon & Schuster, 1995.

Herken, Gregg, *Brotherhood of the Bomb*, New York: Henry Holt and Company, LLC, 2002.

*Hersey, John, *Hiroshima*, New York: Alfred A. Knopf, Inc., 1946.

Hill, D. L., Rabinowitch, E., Simpson, J. A., (Coryell, C. D., English, S., Brown, H., and Sugarman, N.), "The Atomic Scientists Speak Up," *Life* Magazine, October 29, 1945.

Hoddeson, Lillian, P. W. Henriksen, R. A. Meade, and C. Westfall, *Critical Assembly: A Technical History of Los Alamos during the Oppenheimer Years, 1943-1945*, New York: Cambridge University Press, 1993.

Jamison, Kay Redfield, *An Unquiet Mind*, New York: Alfred A. Knopf, Inc., 1995.
_____, *Exuberance, The Passion for Life*, New York: Alfred A. Knopf, 2004.

Johnson, Charles W. and Charles O. Jackson, *City Behind a Fence: Oak Ridge, Tennessee, 1942-1946*, Knoxville: University of Tennessee Press, 1981.

*Johnson, Crockett, *Barnaby,* New York: Henry Holt and Company, 1942, 1943.

*_____, *Barnaby and Mr. O'Malley,* New York: Henry Holt and Company, 1943, 1944.

Joseph, Timothy, *Historic Photos of the Manhattan Project,* Nashville, Tenn.: Turner Publishing Company, 2009.

Kamen, Martin D., *Radiant Science, Dark Politics: A Memoir of the Nuclear Age,* Berkeley: University of California Press, 1985.

Kelly, Cynthia C., ed., *The Manhattan Project: The Birth of the Atomic Bomb in the Words of Its Creators, Eyewitnesses, and Historians,* New York: Black Dog & Levanthal Publishers, Inc., 2007.

*Lang, Daniel, *The Man in the Thick Lead Suit,* New York: Oxford University Press, 1954.

Lanouette, William with Bela Silard, *Genius in the Shadows: A Biography of Leo Szilard, the Man Behind the Bomb,* New York: Scribner's, 1992.

Life's Picture History of World War II, New York: Time Incorporated, 1950.

McMillan, Priscilla J., *The Ruin of J. Robert Oppenheimer and the Birth of the Modern Arms Race,* New York: Penguin Books, 2005.

Selected Bibliography

Nichols, Kenneth D., *The Road to Trinity: A Personal Account of How America's Nuclear Policies Were Made*, New York: William Morrow and Company, Inc., 1987.

Nuclear Detonation Timeline "1945-1998"
http://www.youtube.com/watch?v=I9lquok4Pdk.

Oak Ridge National Laboratory *Review*, Vol. 9, No., 4, Fall 1976, pp.1-62.

Official Register of the United States, U. S. Civil Service Commission, 1953 and 1954.

Oe, Kenzaburo, *Hiroshima Notes*, New York: Marion Boyars Publishers, (1965), 1995.

Palevsky, Mary, *Atomic Fragments: A Daughter's Questions*, Berkeley: University of California Press, 2000.

Rhodes, Richard, *Dark Sun, The Making of the Hydrogen Bomb*, New York: Simon & Schuster, 1995.
_____, *The Making of the Atomic Bomb*, New York: Simon and Schuster, 1985.

Roberts, Arthur, *Physics Songs*.
http://www.haverford.edu/physics/songs/roberts/roberts.htm

Roberts, John D., *The Right Place at the Right Time*, Washington, D.C.: American Chemical Society, 1990.

*Robinson, George O., Jr., *The Oak Ridge Story: The Saga of a People Who Share in History*, Kingsport, Tenn.: Southern Publishers, Inc., 1950.

Schweber, Silvan S., *Einstein and Oppenheimer: The Meaning of Genius*, Cambridge, Harvard University Press, 2008.

**Seaborg, Glenn T., "The First Weighing of Plutonium: Recollections and Reminiscences at the 25th Anniversary," (University of Chicago, September 10, 1967), EOM-303, U. S. Atomic Energy Commission, Computer Registry, 1982.

_____, with Eric Seaborg, *Adventures in the Atomic Age*, New York: Farrar, Straus and Giroux, 2001.

Segrè, Gino, *Faust in Copenhagen, A Struggle for the Soul of Physics*, New York: Viking, 2007.

*Smith, Alice Kimball, *A Peril and a Hope: The Scientists' Movement in America 1945-7*, Chicago: The University of Chicago Press, 1965.

_____, "Behind the Decision to Use the Atomic Bomb: Chicago 1944-45," *Bulletin of the Atomic Scientists*, October 1958.

_____, "The Elusive Dr. Szilard," *Harper's*, July 1960.

*Smyth, Henry DeWolf, *Atomic Energy for Military Purposes. The Official Report on the Development of the Atomic Bomb under the Auspices of the United States Government, 1940-1945*, Princeton: Princeton University Press, 1945.

Steinberg, Ellis P., "Radiochemistry of the Fission Products: Reflections on Nuclear Fission at Its Half-Century,"

Journal of Chemical Education, Volume 66, Number 5, May, 1989, 367-372.

*Tashlin, Frank, *The Bear That Wasn't,* New York: E. P. Dutton, 1946.

Ulam, S. M., *Adventures of a Mathematician,* Berkeley: University of California Press, 1991.

von Hippel, Frank, "James Franck: Science and Conscience," *Physics Today,* June, 2010, 41-46.

Wiesel, Elie, *Rashi,* New York: Nextbook, 2009.

Weizmann, Chaim, *Trial and Error, The Autobiography of Chaim Weizmann,* New York: Harper, 1949.

Wigner, Eugene Paul, *The Recollections of Eugene P. Wigner as told to Andrew Szanton,* New York: Plenum Press, 1992.

"Wyman, Jeffries, 1901-1995," *Biographical Memoirs, Volume 83,* Washington, D. C.: The National Academies Press, 2003.

*Indicates book from Charles D. Coryell's library.
** Indicates Coryell's participation.

Index of Names

About the Editor and Interviewer

Julie Esther Coryell

From childhood, atomic secrets fascinated and frightened Julie Coryell. She chose to study languages, especially Chinese, to differ from both parents. After marriage and graduate school, she and Seelye Martin moved to Seattle and raised two children. Julie taught auto mechanics, and co-founded both Aradia women's health clinic and the Women Studies Program at the University of Washington. Living in Washington DC during International Polar Year revived memories of Charles' public advocacy, friendships, and his *Reminiscences*. In the centennial year of their births, the American Chemical Society honored Glenn Seaborg and Coryell as co-founders of the field of radiochemistry.

Joan Bainbridge Safford

Joan Safford describes her wartime childhood at Los Alamos as "bracketed time." As a graduate student in history, her vivid memories and father's post-war scientists' efforts to control the Bomb, prompted her to initiate interviews for the Columbia University Oral History Research Office during the summer of 1960: her father, Kenneth Bainbridge, Norman Ramsey, and Charles Coryell. Before she could finish editing Charles' tapes, she taught two years in Bogotá, Colombia, raised two children and served as federal prosecutor in Chicago and Justice Department Representative in Mexico, where she continues to consult on criminal justice.

9 780985 671129